Molecular Spectroscopy

John M. Brown

Professor of Chemistry, University of Oxford

Series sponsor: **ZENECA**

ZENECA is a major international company active in four main areas of business: Pharmaceuticals, Agrochemicals and Seeds, Specialty Chemicals, and Biological Products.

ZENECA's skill and innovative ideas in organic chemistry and bioscience create products and services which improve the world's health, nutrition, environment, and quality of life.

ZENECA is committed to the support of education in chemistry and chemical engineering.

OXFORD   NEW YORK   TOKYO
OXFORD UNIVERSITY PRESS
1998

Oxford University Press, Great Clarendon Street, Oxford OX2 6DP

Oxford   New York

Athens   Auckland   Bangkok   Bogota   Bombay   Buenos Aires   Calcutta
Cape Town   Chennai   Dar es Salaam   Delhi   Florence   Hong Kong   Istanbul
Karachi   Kuala Lumpur   Madrid   Melbourne   Mexico City   Mumbai
Nairobi   Paris   São Paolo   Singapore   Taipei   Tokyo   Toronto   Warsaw

and associated companies in
Berlin   Ibadan

Oxford is a trade mark of Oxford University Press

Published in the United States
by Oxford University Press Inc., New York

A catalogue record for this book is available from the British Library

Library of Congress Cataloging-in-Publication Data
Brown, John M.
Molecular spectroscopy / John M. Brown.
(Oxford chemistry primers; 55)
1. Molecular spectroscopy.   I. Title.   II. Series.
QD96.M65B76   1998   543'.0858—dc21   97-46779
ISBN 0 19 855785 X (Pbk)

Typeset by EXPO Holdings, Malaysia
Printed in Great Britain by Arrowhead Books Ltd, Reading

Preface

This book is an introduction to Molecular Spectroscopy, based loosely on a course given to second-year undergraduates at Oxford University. The topic is important for several reasons. Spectroscopy provides a direct and accurate method for the determination of the structure of molecules, their dissociation energies, and ionization potentials. A knowledge of the rotational, vibrational, and electronic levels of molecules is also needed for the calculation of their partition functions, thereby opening up the path to statistical mechanical calculations. Molecular spectra also provide some of the most beautiful and illuminating examples of quantum mechanics in action. Finally, many modern experiments in Physical Chemistry are based on spectroscopy in some form or other. A sound knowledge of the subject is therefore essential if these experiments are to be properly devised and the information determined from them is to be correctly interpreted.

The book is a short one and inevitably there are limits to what it covers. Firstly, it is almost completely restricted to diatomic molecules. No one doubts the significance of this class of molecule but several important aspects of structure and dynamics only manifest themselves in triatomic and larger molecules. Secondly, the book deals with the principles of spectroscopy and not its practice. It should be realized that spectroscopy is above all an experimental subject and the most important work that is done is that carried out in the laboratory. Thirdly, a few important topics have been excluded in order to meet the constraints of length, for example, the spectroscopy of open shell molecules.

Finally, I would like to acknowledge the debt that I owe to several individuals who, in their very different ways, have all contributed to my knowledge of molecular spectroscopy and have opened my eyes to its many delights. They are (in alphabetical order) Philip Bunker, Alan Carrington, Ian Mills, Donald Ramsay, and Jim Watson. I am also very grateful to John Freeman who prepared most of the diagrams in the book.

Oxford   J. M. B.
October 1997

Series Editor's Foreword

Oxford Chemistry Primers are designed to provide clear and concise introductions to a wide range of topics that may be encountered by chemistry students as they progress from the freshman stage through to graduation. The Physical Chemistry series aims to contain books easily recognized as relating to established fundamental core material that all chemists need to know, as well as books reflecting new directions and research trends in the subject, thereby anticipating (and perhaps encouraging) the evolution of modern undergraduate courses.

In this Physical Chemistry Primer, Professor John Brown presents an authoritative and precisely written account of *Molecular Spectroscopy* focusing largely on the spectroscopy of diatomic molecules to make the subject easily accessible to all. The topic is of fundamental importance in physical chemistry and undergraduates will be fascinated by the remarkable level of detail at which we now understand the behaviour of small molecules. This Primer will be of interest to all students of chemistry and their mentors.

Richard G. Compton
Physical and Theoretical Chemistry Laboratory, University of Oxford

Contents

1 Radiation and matter

1.1 Introduction

Towards the end of the Napoleonic wars in Europe, Joseph von Fraunhofer carried out one of the first spectroscopic experiments. He simply focused the direct light from the sun onto a narrow slit, dispersed the light from this image with a prism, and looked carefully at what he saw. (This was effectively the same experiment which Isaac Newton had carried out some 150 years earlier but with much higher dispersion and some ability to measure the wavelengths). In addition to the familiar rainbow pattern formed by spreading out sunlight into its component wavelengths, he saw several black lines superimposed which corresponded to the image of the slit at particular wavelengths. Not knowing the explanation for these features, he simply produced a list of the lines, labelling the major ones alphabetically A, B, C .... H.

It was many years later (1859) that the correct explananation for these lines was suggested by G. R. Kirchoff, namely that the light from the sun was being absorbed at these particular wavelengths by chemical species which are present in the solar atmosphere. The double line in the yellow, labelled D by Fraunhofer, was assigned to the sodium atom because the wavelengths corresponded exactly with the bright lines seen when salt was introduced into the flame of a Bunsen burner. By working systematically through all conceivable possibilities, Kirchoff was able to assign all the Fraunhofer lines to a variety of different elements such as Fe and Ca. This early achievement demonstrates the power of spectroscopy to identify specific atoms and molecules remotely because their spectra are all different and uniquely characteristic. This ability is still exploited today, being used to identify and monitor molecules in remote astrophysical sources like the interstellar clouds, in the further reaches of our own atmosphere, and even in the exhaust gases of cars as they speed past on the highway.

This book is concerned with the basic ideas of molecular spectroscopy. The treatment is confined almost completely to diatomic molecules and the objective is to explain how spectra arise, what they look like in practice and what information about the molecule can be extracted from them. Because spectroscopy is Quantum Mechanics in action, it is necessary to understand something about this subject. Some of its simplest concepts are introduced in a qualitative manner as a framework on which to hang the discussion of the various branches of spectroscopy later on in the book. This dependence of spectroscopy on Quantum Mechanics is direct and inextricable. A thorough study of spectroscopy however returns the favour because it provides an insight into the quantum world from which perception flows.

1.2 The basic spectroscopic experiment

Spectroscopy is the study of the way in which electromagnetic radiation interacts with matter as a function of frequency (or wavelength). The simplest

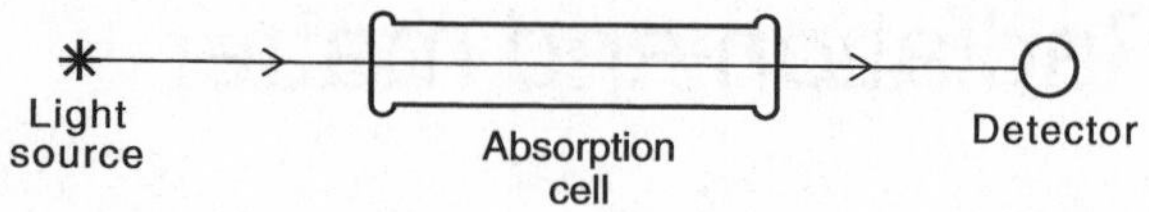

Light source   Absorption cell   Detector

Fig. 1.1   The basic arrangement in a spectroscopic experiment.

possible experimental arrangement is shown in Fig. 1.1. The beam from a tunable source of radiation (nowadays almost always a laser) is passed through through the sample contained in a cell and its intensity is measured by a detector. The variation of the signal intensity as the frequency of the radiation is scanned is called the *spectrum*. If the light wave does not interact at all with the sample in this frequency range, the spectrum is flat and uninformative (the sample is said to be transparent). If, however, the radiation interacts at particular wavelengths, energy is absorbed and various peaks occur in the spectrum (corresponding to a reduction in intensity at the detector). These absorption features are often called 'lines', harking back to the very earliest spectra when lines were indeed recorded, being the image of a slit on a photographic plate (or on the retina of the eye in the case of Fraunhofer). The spectral wavelengths recorded in this way are characteristic of the molecule being studied and are a primary source of information on its structure and dynamics.

1.3   Electromagnetic radiation

Electromagnetic radiation is described as a *transverse waveform*. By this is meant that it consists of oscillating electric and magnetic fields which point transversely to the direction of propagation of the wave. For example, for linearly polarized light, the oscillating electric field might point in the X direction, the oscillating magnetic field in the Y direction, and Z is the direction of propagation, see Fig. 1.2. The plane of polarization is conventionally defined as the plane containing the E field and the direction of propagation (XZ). Both the electric and magnetic field are oscillating at the same frequency ν (in units of s^{-1} or Hz) and the light wave travels through a vacuum at a very high (but finite) speed, $c = 2.9979 \times 10^8$ m s^{-1}. The distance between adjacent crests at a given point in time is called the wavelength, λ:

$$\lambda = c/\nu. \tag{1.1}$$

What is most important from the point of view of spectroscopy is that energy can be transferred from, say, the source to the detector in the form of electromagnetic radiation. If the instantaneous electric and magnetic field strengths are E and H, respectively, the energy density (that is the energy stored in unit volume) is

$$W/\mathrm{J\,m^{-3}} = \tfrac{1}{2}\varepsilon_0 E^2 + \tfrac{1}{2}\mu_0 H^2 \tag{1.2}$$

where ε_0 is the permittivity and μ_0 is the permeability of a vacuum. The energy is stored equally in the electric and magnetic fields. The intensity I of this light wave is the energy crossing unit area per second

$$I/\mathrm{J\,s\,m^{-2}} = \tfrac{1}{2}\varepsilon_0 c E_0^2 = \tfrac{1}{2}\mu_0 c H_0^2 \tag{1.3}$$

where E_0 and H_0 are the amplitudes of the electric and magnetic sinusoidal waveforms (that is, the maximum electric and magnetic fields).

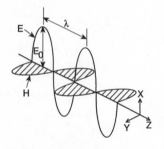

Fig. 1.2   A plane electromagnetic wave, seen at an instant in time and propagating in the Z direction.

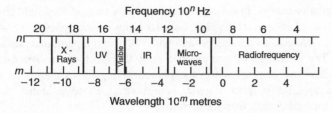

Fig. 1.3 The various regions of the electromagnetic spectrum, with their defining frequencies and wavelengths, plotted on a logarithmic scale.

Although electromagnetic radiation has the same mathematical description throughout its whole spectrum, we are used to thinking of it in separate sections because of the different effects it is has on its surroundings. The section which is most familiar is of course the region of visible light, spreading out from red light at the long wavelength end to violet at the short. Moving out to longer wavelengths (or lower frequencies) from the visible, we come in succession to the infrared, the microwave and finally the radiofrequency regions. On the other side of the visible region, we encounter the ultraviolet and then the X-ray regions (beyond this lie the regions of γ-rays and cosmic rays but radiation at these wavelengths is not used in laboratory-based spectroscopy). The wavelengths and frequencies which define these regions are shown in Fig. 1.3.

Essentially all spectroscopic effects can be described by treating the radiation classically, as a transverse waveform. It is sometimes convenient and simpler to describe light of frequency v from a quantum standpoint. From this perspective, light is regarded as a stream of particles, called *photons*, each of which carries an energy hv, where h is Planck's constant (of which, more anon) and v is the frequency. This was Einstein's explanation for perplexing observations such as the photoelectric effect which could not be understood on the classical basis that light consisted of a continuous electromagnetic field. In order to explain some of the phenomena associated with light, it was necessary to ascribe rather bizarre properties to the photon. Thus, although a photon has no mass, it does possess linear momentum of magnitude hv/c. It also carries one unit of angular momentum ($\pm h/2\pi$).

1.4 The interaction between radiation and matter

Spectroscopy is the study of the exchange of energy between radiation and matter. The mechanism by which this occurs involves the interaction between the oscillating electric (or magnetic) field in the radiation with the appropriate dipole moment in the molecule. In order for this interaction to be strong, the dipole moment must be oscillating in some way (for example, through rotation or vibration of the molecule) at the same frequency as the radiation field. The spectrum recorded by Fraunhofer of material on the sun is quite typical of all spectroscopic observations in that matter can only respond to light at particular, discrete frequencies. There is no absorption of light in between individual lines. This is a direct effect of the quantization of energy at the molecular level. As we shall see later in this book, at the scale of atoms and molecules, the energy of the system cannot take on any value but only those which correspond to so-called *stationary states*. This behaviour is called the quantization of energy. In exactly the same way as we described light above as a collection of individual photons each carrying its own energy, so we can think of a molecule as moving around with only certain permitted amounts of energy. The exchange of energy that is at the heart of the spectroscopic process therefore corresponds to the photon giving up its energy $h\nu$ to the molecule to raise it from a lower energy level E_1 to a higher one E_2. Conservation of energy therefore requires

$$h\nu = E_2 - E_1. \tag{1.4}$$

If energy is emitted in the form of light (i.e. the reverse process takes place), eqn 1.4 still holds good. The molecule falls to a lower level and emits a photon of the appropriate frequency. Equation 1.4 enshrines an important statement, due primarily to Einstein and Planck, which says that the energy change produced in a molecule is proportional to the frequency of the radiation responsible. The constant of proportionality is called Planck's constant and has a value of 6.6261×10^{-34} J s, in other words, it is very small.

1.5   Relative energy units in spectroscopy

The practice of spectroscopy consists of the measurement of the discrete amounts of energy which are passed between molecules and radiation when they interact. These energy exchanges can then be interpreted in terms of the structural properties of the molecule as we shall see in later chapters. In the laboratory however, it is not energy that is measured directly. Rather, in the radiofrequency, microwave, and far-infrared regions, spectral measurements are made in frequency units (in Hz). For infrared, visible, and ultraviolet spectroscopy where the primary standard of measurement is length, it is the wavelength λ (in m) of the radiation which is measured. It might seem a simple exercise to convert these quantities to energy units by invoking eqn 1.4 or equivalently

$$hc / \lambda = E_2 - E_1. \tag{1.5}$$

However, this conversion is almost never made in practice, partly because it is not necessary but more importantly because it is often the case that the frequency of a transition can be measured much more accurately than the uncertainty with which Planck's constant h is known. (It used to be necessary to consider the uncertainty in the speed of light c also but this quantity was defined to be exactly 299792458 m s^{-1} in 1983.) Therefore the amount of energy is less reliably known than the frequency from which it is derived. Consequently, the energy change involved in spectroscopic transitions is measured in relative units, either *frequency* in Hz, MHz, GHz (or whatever is appropriate) or the inverse of the wavelength (called the *wavenumber*) in m^{-1} or more commonly cm^{-1}. The wavenumber $\tilde{\nu}$ is defined by

$$\tilde{\nu} = 1 / \lambda_{vac} = 1 / (\lambda_{air} n_{air}) \tag{1.6}$$

where n_{air} is the refractive index of air for the wavelength concerned. This conversion is relevant because many spectrometers operate with their optical paths through air at atmospheric pressure.

Mention should also be made of another unit of energy which is useful in spectroscopy, namely the electron-volt (or eV). This is the amount of energy

Table 1.1 Conversion factors of energy units

	cm^{-1}	GHz	eV	E/aJ^a
cm^{-1}	1	29.97925	1.2398×10^{-4}	1.9864×10^{-5}
GHz	3.3356×10^{-2}	1	4.1357×10^{-6}	6.6261×10^{-7}
eV	8065.54	2.41799×10^5	1	0.16022
E/aJ^a	50341.1	1.50919×10^6	6.24151	1

[a] 1 aJ (or attojoule) is 10^{-18} joules.

which an electron acquires when it has been accelerated through a potential of 1 V. Substituting the charge of an electron in Coulombs, we calculate that 1 eV corresponds to 1.6022×10^{-19} J molecule^{-1}. This unit is particularly appropriate to the measurement of ionization processes but it is also useful for the measurement of separations between electronic states of molecules. These separations correspond to a large number of cm^{-1} (several tens of thousands) but only a few eV. The conversion factors between these various units are summarized in Table 1.1.

The equilibrium distribution of molecules among a set of energy levels depends on their average thermal energy kT. In order to assess the population distribution readily, it is worthwhile to commit this quantity in relative energy units to memory (kT at 298 K is 207.1 cm^{-1}, in round figures 200 cm^{-1}).

2 Quantization and molecular energy levels

2.1 Introduction

Physical properties of atoms and molecules are *quantized*, that is they can only take certain discrete values rather than a full range of continuously varying values. This is in contrast to everyday effects which we observe in the world around us and which are well described within the framework of classical physics such as Newton's Laws. It is hard to imagine, for example, being able only to throw a ball through the air at certain velocities yet this is what happens in a quantum world. As a result, it is a common misconception that the laws of classical and quantum physics are in conflict. This is not the case. Rather, the laws of quantum mechanics go smoothly over to those of classical physics as the scale of the system increases. Newton's Laws are the large scale limit of the Schrödinger equation. You could describe the flight of a ball through the air by solving Schrödinger's equation but it is much easier to use Newton's Laws! The point on the size scale at which quantum laws go over to classical laws depends on the size of Planck's constant h (which equals $6.626\,076 \times 10^{-34}$ J s). If this constant were much larger, say of the order of 10^{-10} J s, we would be experiencing quantum effects all the time. This connection between small and large scale behaviour was recognized early on by the physicists working in this area. It was called the Correspondence Principle and played an important rôle in guiding them to the laws of Quantum Mechanics.

2.2 The Schrödinger equation

At the quantum level, a system is not described by the values of its properties such as velocity, momentum, or energy but rather by a wavefunction which in general is a function of spatial and temporal co-ordinates, $\Psi(x, y, z; t)$. The classical, measurable properties are represented by *operators* which act on the wavefunctions. For example, a position co-ordinate is represented simply by itself:

$$x \to \hat{\mathbf{x}} \tag{2.1}$$

whereas the component of linear momentum is represented by a complex, partial derivative:

$$p_x \to \hat{\mathbf{p}}_X = -i\,\hbar\,\partial/\partial x, \tag{2.2}$$

where $\hbar = h/2\pi$. (It is a common practice to place 'hats' ˆ on the symbols to emphasize the fact that they are operators. With a little experience, this property becomes obvious from the context and the 'hats' are discarded.) The value obtained for a property by measurement is given by its *expectation value*. For example, a measurement of the x-component of the linear momentum operator gives a value for the quantity

$$\int \Psi^*(-i\hbar\partial/\partial x)\Psi d\tau \qquad (2.3)$$

where the integral is actually a multiple integral and $d\tau$ represents the differential volume element for all the appropriate coordinates.

In spectroscopy, we are particularly interested in the quantized energy levels of a molecule. These are the so-called *stationary states* whose energy is independent of time. In this case, the energy of the system is represented by a more complicated operator called the *Hamiltonian* (after the Irish mathematician, William Hamilton, 1805–1865, who developed a formulation of Newton's equations in terms of position coordinates and momenta). The procedure for constructing a quantum mechanical Hamiltonian is quite straightforward:

(1) write down the complete, classical expression for the energy of the system (for example, the sum of the kinetic and potential energies);

(2) re-express this result, if necessary, in terms of linear momenta and position coordinates; and

(3) use the recipes given in eqns 2.1 and 2.2 above to transform the classical function into operator form. The result is the Hamiltonian operator.

(This procedure works for the construction of any quantum mechanical operator and is in fact the unjustified formulation of wave mechanics by Schrödinger and his colleagues.)

As a simple example of this procedure, let us consider the formulation of the Hamiltonian for a single particle of mass m moving within the confines of a potential well. The classical energy is the sum of the kinetic energy T and the potential energy V:

$$E_{class} = T + V \qquad (2.4)$$

$$\text{where} \qquad T = \tfrac{1}{2} m (\dot{x}^2 + \dot{y}^2 + \dot{z}^2) \qquad (2.5)$$

$$\text{and} \qquad V = V(x, y, z). \qquad (2.6)$$

Here (x, y, z) are the coordinates of the particle and $\dot{x}$ is its velocity component in the x direction, dx/dt. The potential energy V is an unspecified function of the coordinates. For example, if $V = 0$, we are describing the motion of a free particle. In the next stage, we convert eqn 2.5 to linear momenta, using for example

$$p_x = m \dot{x}. \qquad (2.7)$$

Hence

$$T = 1/2m (p_x^2 + p_y^2 + p_z^2). \qquad (2.8)$$

To transform this result to the quantum mechanical form, we need to deal with terms such as p_x^2:

$$p_x^2 = p_x p_x \rightarrow -i\hbar\, \partial/\partial x\,(-i\hbar\, \partial/\partial x) = -\hbar^2 \partial^2/\partial x^2. \qquad (2.9)$$

Using the corresponding transformation for p_y^2 and p_z^2 and substituting in eqn 2.4, we obtain the expression for the Hamiltonian operator

$$H = -\hbar^2/2m\,(\partial^2/\partial x^2 + \partial^2/\partial y^2 + \partial^2/\partial z^2) + V(x, y, z). \qquad (2.10)$$

The first term in parentheses on the right hand side occurs often in physics and chemistry. It is known as the Laplacian operator and given a symbol ∇^2 (del squared). In quantum mechanics, we get used to this operator being called upon to represent the kinetic energy.

The time independent properties of our single particle are described by its wavefunction $\Psi(x, y, z)$. This is obtained by solving the Schrödinger equation:

$$H\Psi = E\Psi \qquad (2.11)$$

where E is the energy of the quantum system. This equation has been written in the form of an *eigenvalue equation*; the effect of the operator H acting on the wavefunction is identical to simply multiplying the same wavefunction by a number (i.e. scaling it). In this situation, Ψ is an eigenfunction of H and E is its corresponding eigenvalue. Integration of eqn 2.11 yields:

$$\int \Psi^* H \Psi \, d\tau = E; \qquad (2.12)$$

in other words, as we anticipated, the energy of the state is given by the expectation value of H. In general, a quantum mechanical operator such as H will have a whole set of eigenfunctions and corresponding eigenvalues, each of which satisfies the Schrödinger equation. The process of 'solving' the Schrödinger equation is the determination of these eigenfunctions and eigenvalues.

The general approach to the determination of the energy levels of a physical system involves two stages:

1. The definition of the Hamiltonian appropriate to the system. With enough commitment, this can always be done rigorously although the resulting operator may be very complicated.

2. The determination of the eigenfunctions and eigenvalues of this Hamiltonian. In practice, this is an exercise in the solution of second order, partial differential equations.

The Schrödinger equation (eqn 2.11), can only be solved exactly for one-body systems such as the particle in a box or the simple harmonic oscillator. Almost all problems of interest involve several particles; for these, one has to resort to approximate methods of solution (e.g. numerical methods, perturbation theory).

The quantization of energy levels usually arises from the imposition of boundary conditions on the wavefunction. For example, for a particle moving in a one-dimensional box with infinitely high walls, acceptable solutions must go to zero at the boundaries. Any wavefunction which does not satisfy this condition might correspond to an intermediate energy value but would not describe a stationary state of the system. If the particle is translating freely, (V is zero in eqn 2.10), there are no boundaries or boundary conditions and all possible translational energies are allowed.

2.3   Molecular wavefunctions: Born–Oppenheimer separation

We are now in a position to be able to construct the Hamiltonian operator for an individual molecule. We consider the molecule to be a collection of charged, massive particles (electrons and nuclei) which move under the influence of electrostatic forces. The instantaneous position of each of n such particles can be specified by three Cartesian coordinates; the system is thus described by a set of $3n$ coordinates. The construction of the Hamiltonian in terms of these coordinates and their conjugate momenta is comparatively straightforward, using the procedure outlined in the previous section. The solution of the resultant Schrödinger equation, on the other hand, is much more difficult. Modern *ab initio* methods have been applied to this problem with some success; generally speaking, the calculations become more reliable as the power of the computer increases. However, a really accurate calculation is still only possible for a molecule containing very few electrons, something like the CH molecule with seven electrons, for example. Calculations are carried out on much larger molecular systems of course but with increasing resort to approximations. Fortunately, if we are interested in only certain aspects of a molecule's energy levels, there is a very useful approximation which is reliable in most circumstances. This approximation is based on the hierarchy of energy levels and is known as the Born–Oppenheimer separation after its two original proponents, M. Born and J. R. Oppenheimer.

The physical basis for the Born–Oppenheimer separation is as follows: In a molecule, the electrons and nuclei of its component atoms are subjected to forces of similar magnitude (these forces are electrostatic in origin and the interaction is mutual). However, since the nuclei are about four orders of magnitude more massive, the electrons move much more rapidly than the nuclei. To a good approximation, the problem of the motion of the electrons can therefore be treated as if the nuclei were fixed; this problem is simpler to solve (though still quite complicated) because the number of coordinates has been reduced. Mathematically speaking, this approximation amounts to a factorization of the total wavefunction into an electronic and a nuclear part; the electronic wavefunction is a function of electronic coordinates only (for fixed internuclear separations):

$$\Psi_{tot} = \Psi_{el}(q_{el})\Psi_{nucl}(q_{nucl}). \tag{2.13}$$

In the same spirit, the nuclear motion can be further separated into a vibrational and a rotational part. Vibrational motion for a diatomic molecule is defined to be the variation in the internuclear separation whereas rotational motion is the change in orientation of the molecule in laboratory-fixed space. Empirical observation shows that the separation between vibrational levels is much larger than that between rotational levels, i.e. the characteristic vibrational frequencies are larger than the rotational frequencies. Therefore, we can describe the vibrational motion within the averaged potential energy created by the rapidly moving electrons for a fixed molecular orientation:

$$\Psi_{tot} \approx \Psi_{el}(q_{el})\Psi_{vib}(q_{vib})\Psi_{rot}(q_{rot}). \tag{2.14}$$

To the extent that this wavefunction adequately describes the physical situation, we have succeeded in solving a second order, partial differential equation by separation of variables into electronic, vibrational, and rotational sets. Consequently, we can write the total energy as a simple sum of contributions from these three types of motion:

$$E_{tot} = E_{el} + E_{vib} + E_{rot}. \tag{2.15}$$

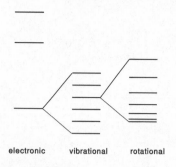

electronic   vibrational   rotational

Fig. 2.1 The arrangement of energy levels of a diatomic molecule, showing the Born–Oppenheimer classification. Each electronic state supports a set of vibrational energy levels; each vibrational level in turn has its own set of rotational levels.

Although the Born–Oppenheimer separation is only an approximation, it is in fact a very good approximation in the vast majority of situations. Very accurate measurements are required to detect the breakdown of the electronic Born–Oppenheimer separation. It is of enormous help that such a simple form as that in eqn 2.14 gives a reliable description of the energy level scheme. The familar hierarchy of energy levels associated with it ($\Delta E_{el} \gg \Delta E_{vib} \gg \Delta E_{rot}$) is depicted in Fig. 2.1.

No mention has been made of another familiar type of molecular motion, namely translation. There are two reasons for this. Firstly, in the absence of external electric or magnetic fields, there is a rigorous separation of the translational motion of the molecule as a whole from its other degrees of freedom. This separation is achieved by referring all the other coordinates (q_{el}, q_{vib}, q_{rot}) to the molecular centre of mass rather than to some laboratory-fixed origin. The three coordinates (X, Y, Z) which define the instantaneous position of the centre of mass are then the translational coordinates. The second reason for not including translational motion in our description is that spectroscopy studies transitions which occur *within* individual molecules, each of which is moving at a translational velocity **v**. The only effect that translational motion has on the observed transition frequency is a small, indirect one, through the Doppler effect.

The behaviour of quantum systems can be markedly different from their classical analogues. This difference in behaviour becomes more pronounced the larger the size of the quantum. Thus a classical, Newtonian description of translational motion is completely adequate for atomic or molecular systems. Even for rotational motion, a classical picture is quite reliable and helpful. However, for electronic motion, the description as a charged particle moving in an electrostatic field provides almost no insight at all. This result is related to the Correspondence Principle mentioned earlier, which states that quantum systems go over to classical behaviour in the limit of large quantum numbers.

2.4 Pauli exclusion principle: nuclear statistical weights

At a molecular level, a system can only be described by a wavefunction. Consequently, it is impossible to distinguish between identical particles such as the electrons or nuclei in a molecule. Thus molecules which contain such identical particles (as all do in the form of electrons) possess a symmetry associated with this property, which can be used to characterize their wavefunctions.

The permutation of identical particles: the exclusion principle

Let P_{12} be the operator which interchanges a pair of identical particles 1 and 2 or permutes their labels, which amounts to the same thing. Under this operation, a general function of the coordinates of these particles transforms as follows:

$$P_{12}f(x_1; x_2) = f(x_2; x_1); \tag{2.16}$$

that is, particle 1 now has the coordinates of particle 2 and vice versa. If the two particles are identical, the states of the system obtained simply by interchanging them in this way must be completely equivalent; in other words, the effect of P_{12} on the system leaves it invariant. However, the effect on the wavefunction is to transform it into either itself or the negative of itself:

$$P_{12}\Psi(x_1; x_2) = \pm\Psi(x_2; x_1). \qquad (2.17)$$

These two possibilities are allowed because the properties of the system always depend on the product of two wavefunctions (usually the square). It is an empirical observation of quantum physics that the upper sign applies when the particles obey Bose–Einstein statistics whereas the lower sign choice applies when they obey Fermi–Dirac statistics. In the former case, the particles are called *bosons* and have total spin = 0, 1, 2, 3, ... In the latter case, the particles are known as *fermions* and have half-integral spin (1/2, 3/2, 5/2, ...). Thus for bosons, the wavefunction must be symmetric with respect to permutation of the two particles whereas for fermions it is antisymmetric. The electron has a spin of $\frac{1}{2}$ and is therefore a fermion. As a result, molecular wavefunctions must be antisymmetric with respect to interchange of pairs of electrons. This requirement is equivalent to the simpler statement of the Pauli exclusion principle that no two electrons can have the same set of quantum numbers. If any pair did have the same quantum numbers, the wavefunction would be unaffected (i.e. it would be symmetric) with respect to interchange of these two electrons. The implications of the Pauli exclusion principle for electronic states will be discussed in Chapter 7.

Permutation of identical nuclei: nuclear statistical weights

In a homonuclear diatomic molecule, the two nuclei are also indistinguishable. In this case, the exclusion principle imposes constraints on the nuclear wavefunctions as a result of which only some of them are allowed. Consider, for example, the case of molecular hydrogen, H_2. The nucleus of the H atom is simply a proton and has a spin $I = \frac{1}{2}$. It is a fermion and so the total wavefunction must be antisymmetric with respect to the permutation of the two nuclei, which are labelled a and b:

$$P_{ab}\Psi_{tot} = -\Psi_{tot}. \qquad (2.18)$$

Following on from the Born–Oppenheimer separation, we shall adopt the following (approximate) form for the wavefunction:

$$\Psi_{tot} = \Psi_{el}\Psi_{vib}\Psi_{rot}\Psi_{ns} \qquad (2.19)$$

where Ψ_{ns} is the nuclear spin wavefunction. We can classify the total wavefunction by investigating the effect of P_{ab} on each of the factors on the right hand side.

We deal with the electronic factor first. The wavefunction Ψ_{el} is a function of the electronic coordinates (x_i, y_i, z_i), $i = 1, N$, which are conveniently defined in a local axis system with its origin at the molecular centre of mass; this coordinate system is known as the *molecule-fixed axis system*. It might be thought that the simple action of swapping the nuclear labels a and b would have no effect at all on the electronic coordinates. This is not strictly so because the molecule-fixed axis system must be attached to the molecule according to some prescription. Let us stipulate, for example, that the z axis lies along the internuclear axis with its positive direction running from a to b; the x and y axes will therefore lie perpendicular to the molecular axis. After applying the operator P_{ab} to this system, we find that the z axis is now pointing in exactly the opposite direction, see Fig. 2.2. This is equivalent to saying that all the electrons have been rotated through 180° relative to the molecule-fixed

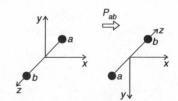

Fig. 2.2 The effect of permuting two identical nuclei a and b on the orientation of the molecule-fixed axis system.

axis system by this operation, even though they will not have moved at all when viewed from the laboratory-fixed axis system. If our arrangement of the molecule-fixed axis system means that P_{ab} causes a rotation of $180°$ about the x axis, then

$$P_{ab}\, x_i = x_i, \tag{2.20a}$$

$$P_{ab}\, y_i = -y_i, \tag{2.20b}$$

$$P_{ab}\, z_i = -z_i. \tag{2.20c}$$

Hydrogen has a closed shell structure in its ground state; its electronic wavefunction has Σ_g^+ symmetry. Such a function is unaffected by reversing the signs of the y_i and z_i coodinates and so

$$P_{ab}\Psi_{el} = +\Psi_{el}. \tag{2.21}$$

Next we consider the vibrational factor, Ψ_{vib}. Since the vibrational motion is the variation in the bond length R of the diatomic molecule, we can use this quantity as the vibrational coordinate. (A more discriminating choice might be the variation in the bond length from its equilibrium value, R_e, but the actual choice does not affect the present argument.) It is easy to see that swapping over the labels on the two nuclei has no effect at all on the coordinate R, i.e. it is symmetric with respect to P_{ab}. Therefore any function of R, in particular the vibrational wavefunction Ψ_{vib}, is also symmetric with respect to P_{ab}:

$$P_{ab}\Psi_{vib} = +\Psi_{vib}. \tag{2.22}$$

We now turn to the rotational wavefunction. The rotational coordinates are the angles needed to define the orientation of the molecule-fixed axis system with respect to the laboratory-fixed system. In the general situation, three such angles are required (the Euler angles) although, for the special case of the diatomic molecule, only two are necessary since the molecule has cylindrical symmetry about the z axis. Again, it might be thought that the simple interchange of the labels on the two nuclei would have no effect on these rotational coordinates since clearly the orientation of the molecule in laboratory space is unaltered. However, we need to remember that the molecule-fixed axis system changes the way it points when a and b are swapped over, see Fig. 2.2. Consequently, in order to determine how Ψ_{rot} transforms under P_{ab}, we need to know how it transforms under a rotation through $180°$ about the x axis, see eqn 2.20. We shall see in Chapter 4 that the rotational wavefunctions of a diatomic molecule are simply the spherical harmonics, which also form the angular wavefunctions for the hydrogen atom. Thus the $J = 0$ eigenfunction has the same angular form as the s orbital, $J = 1$ as the p orbital, $J = 2$ as the d orbital, and so on. From the form of these functions, we know that those with even J are unaffected by a rotation through $180°$ whereas those with odd J change sign; this is indicated in Fig. 2.3. Thus we have:

$$P_{ab}\Psi_{rot,J} = (-1)^J \Psi_{rot,J}. \tag{2.23}$$

In other words, the rotational factor is symmetric for even J and antisymmetric for odd J.

Finally, we consider the nuclear spin wavefunction, Ψ_{ns}. For each H nucleus, there are two possible spin states, α and β which correspond to $M_I = +\frac{1}{2}$ and $-\frac{1}{2}$. For the two spin system in H_2, we therefore have four possible

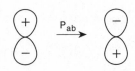

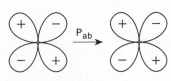

Fig. 2.3 The effect of the nuclear permutation operator P_{ab} on the rotational wavefunctions for $J = 1$ and 2.

spin states $\alpha\alpha$, $\alpha\beta$, $\beta\alpha$, and $\beta\beta$ where the first factor in the product refers to nucleus a and the second to nucleus b. Taking appropriate combinations, it is easy to show that three of these functions are symmetric with respect to P_{ab}:

$$P_{ab}\alpha\alpha = \alpha\alpha \tag{2.24a}$$

$$P_{ab}1/\sqrt{2}(\alpha\beta + \beta\alpha) = 1/\sqrt{2}(\alpha\beta + \beta\alpha) \tag{2.24b}$$

$$P_{ab}\beta\beta = \beta\beta \tag{2.24c}$$

whereas the fourth combination is antisymmetric:

$$P_{ab}1/\sqrt{2}(\alpha\beta - \beta\alpha) = -1/\sqrt{2}(\alpha\beta - \beta\alpha). \tag{2.25}$$

The three symmetric combinations form the three components of a *triplet* spin state (one with a total nuclear spin $I_T = 1$) whereas the antisymmetric combination is a *singlet* on its own (with $I_T = 0$).

We are now in a position to combine all these results to consider the transformation of the total wavefunction. The exclusion principle requires this wavefunction to be antisymmetric with respect to P_{ab} (eqn 2.18). Both Ψ_{el} and Ψ_{vib} are symmetric and so do not affect the overall result. To obtain antisymmetric behaviour, we need to combine even-J rotational levels with the antisymmetric spin combination, the singlet, or to combine the odd-J rotational wavefunction with the symmetric, triplet nuclear spin function. Because the hyperfine splittings between the different nuclear spin states are very small, they are not usually resolved. Consequently, the even-J levels are non-degenerate from the point of view of their hyperfine structure while the odd-J levels appear to have a degeneracy of three. These factors are referred to as *nuclear statistical weights*; the levels with higher weight are known as *ortho* and those with lower weight are known as *para*. (This association can be remembered by reference to the words *ortho*dox and *para*dox.) The first few rotational levels of H_2 are shown in Fig. 2.4.

We see from the above that, for even J levels, three quarters of the possible hyperfine states are missing whereas for odd J levels, one quarter is missing. In the high temperature limit, when many rotational levels are populated, on average $\frac{1}{2}$ $(\frac{3}{4} + \frac{1}{4})$ or $\frac{1}{2}$ of the levels are missing. The inverse of this factor is sometimes known as the symmetry number. In the generalization of this result, for a homonuclear diatomic molecule with nuclei of spin I, of the possible total of $(2I + 1)^2$ states, $(I + 1)(2I + 1)$ hyperfine states are allowed for *ortho* levels and $I(2I + 1)$ are allowed for *para* levels. The *ortho* to *para* ratio is therefore $(I + 1)/I$. For $I = 0$, this means that half the rotational levels (those with odd J for $^1\Sigma_g^+$ states) are not allowed at all.

Transfer from an *ortho* state to a *para* state (or vice versa) requires a re-coupling of the individual nuclear spins. This is very unlikely to occur in practice and, for a homonuclear diatomic molecule, *ortho* and *para* states can be considered in many respects as separate species. Transitions between *ortho* and *para* forms of diatomic molecules, whether induced spectroscopically or by collisions, are therefore highly forbidden.

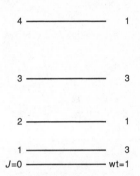

Fig. 2.4 The first few rotational levels of H_2 in its electronic ground state, showing the alternating *para* and *ortho* nuclear statistical weights.

3 Transition probabilities and selection rules

3.1 Spectroscopic transitions

A spectroscopic experiment registers the change of a molecule from one quantum state to another; this process is known as a *spectroscopic transition*. The energy required to drive such a process is usually provided by electromagnetic radiation. At the heart of the spectroscopic experiment therefore is the exchange of energy between radiation and matter.

In the previous chapter, we considered the description of the stationary states of a quantum system, using the time-independent Schrödinger equation. The change of state brought about in a spectroscopic transition requires us to solve the time-dependent Schrödinger equation, which is somewhat more complicated:

$$H \Psi = i \hbar \partial \Psi / \partial t; \tag{3.1}$$

here t is the time coordinate. Fortunately, if we are dealing with weak oscillating electromagnetic fields, we can use time-dependent perturbation theory to describe the spectroscopic transition. This theory provides a simple, systematic approach to the solution of the problem. The reader is referred to a quantum mechanics textbook for a more complete discussion of this topic.

The transitions studied in molecular spectroscopy usually involve the interaction between the electric dipole moment of the molecule, μ, and the electric field of the radiation, $\mathbf{E}$. The interaction energy W_{el} can be expressed very simply by:

$$W_{el} = -\boldsymbol{\mu} \cdot \mathbf{E}. \tag{3.2}$$

There is an orientational dependence in this interaction because the quantities involved are vectors. In converting the energy expression to quantum mechanical form, it is possible to treat the field classically and the dipole moment quantum mechanically. This procedure is justified when the radiation is sufficiently intense that the appearance or disappearance of a single photon has a negligible effect on the intensity. The dipole moment operator

$$\boldsymbol{\mu} = \sum_i q_i \mathbf{r}_i \tag{3.3}$$

is simple to convert to quantum form using the procedure given in Section 2.2 since it only involves positional operators. We shall return to this topic in more detail later.

The molecular dipole moment in eqn 3.3 varies in time as a result of the motion of the molecule. This variation is slow for low frequency motions (such as rotations) and rapid for high frequency motions (such as electronic). For the interaction between the electromagnetic radiation and the molecule to be strong, there is a 'resonance' condition which must be satisfied. Classically, this amounts to the electric dipole moment oscillating at the same frequency as the oscillating electric field of the radiation. If these two motions are in phase, the exchange of energy is most likely to occur. Quantum mechanically, this condition is expressed as:

$$h\nu = E_u - E_l \qquad (3.4)$$

where ν is the frequency of the radiation and E_u, E_l are the upper and lower energy levels.

Electric dipole transitions are much the most common in molecular spectroscopy. However, if we remember that a light wave consists of related oscillating electric and magnetic fields, it is pertinent to ask if the transfer of energy can be brought about by a magnetic interaction instead. In this case, the interaction is described by

$$W_{\mathrm{mag}} = -\mathbf{m} \cdot \mathbf{B} \qquad (3.5)$$

where $\mathbf{m}$ is now the magnetic dipole moment and $\mathbf{B}$ is the magnetic flux density. To observe transitions in this way, we simply require the molecule to possess a magnetic dipole moment which oscillates in time with the motion to be studied. Since a molecule is a collection of charged particles (electrons and nuclei) in motion, it will indeed possess such a magnetic moment. However, the magnitude of the magnetic moments which occur in practice are small, and the transitions induced by magnetic interactions are typically 10^4 times less likely to occur than the corresponding electric dipole transitions. In a situation where both can occur, the electric dipole transition dominates. It is only for systems which do not possess an oscillating electric dipole moment (such as electron spin or nuclear spin) that magnetic dipole transitions are routinely observed in practice (in ESR or NMR experiments).

3.2 Radiative relaxation; Einstein A and B coefficients

Let us consider a pair of isolated, molecular energy levels, labelled 1 and 2. Although there are several ways in which molecules can be persuaded to move between these levels, only two of them involve radiation, namely, spontaneous emission and induced emission (or absorption). These two processes are shown in Fig. 3.1.

Spontaneous emission

If a molecule is in the upper level 2, it will tend to lose its energy by relaxation to level 1, giving out a photon of energy $h\nu_{21}$ in the process. Einstein showed that the probability of this occurring for a single molecule is given by

$$A/\mathrm{s}^{-1} = 16\pi^3 \nu_{21}{}^3 |\mu_{12}|^2 /(3\varepsilon_0 h c^3). \qquad (3.6)$$

In this expression, ν_{21} is the frequency (in Hz) of the radiation emitted, ε_0 is the permittivity of free space, and μ_{12} is the transition dipole moment (in C m). This last quantity is the same as the oscillating dipole moment in our discussion above. It is probably difficult to see why such unforced emission of energy, which occurs whether radiation of the appropriate frequency is present or not, needs to involve the transition moment. This fact is only properly appreciated when the radiation field is also treated in a quantum fashion. In such a treatment (quantum electrodynamics), spontaneous emission occurs when the molecule interacts with the radiation vacuum state, which is something akin to the zero point energy state for quantized radiation.

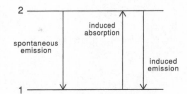

Fig. 3.1 The various radiative transition processes which can occur between a pair of energy levels in a molecule. Spontaneous emission occurs in the absence of external radiation; the two induced processes require the presence of an electromagnetic radiation field of the appropriate frequency.

Each time a molecule undergoes spontaneous emission, it gives out an amount $h\nu_{12}$ of energy. Thus, if there are N_2 molecules in level 2, energy is emitted at the rate

$$I_{21} \; (\mathrm{J\,s^{-1}}) = N_2 h\nu_{12} A. \tag{3.7}$$

Combining 3.6 and 3.7, we see that the intensity of emitted light depends on the fourth power of the frequency. Thus, spontaneous emission becomes an increasingly important decay process as the frequency increases. At microwave frequencies, it is almost irrelevant ($A = 3 \times 10^{-7}$ s^{-1} for $\nu = 3$ GHz), whereas in the ultraviolet, it is often the dominant relaxation mechanism and sets a limit to the lifetime of the molecule in the upper state (for $\lambda = 250$ nm, $A = 2 \times 10^7$ s^{-1}).

Induced emission and absorption

Induced emission and absorption are brought about by a coupling between the oscillating transition moment and the oscillating electromagnetic field. The stronger the field, the more likely the transition is to occur. Einstein showed that the probability of induced absorption for a single molecule is given by

$$P_{12} = \rho(\nu_{12}) B_{12} \tag{3.8}$$

where $\rho(\nu)$ is the radiation density (in J m^{-3}) at a frequency ν, see eqn 1.2, and B_{12} is the Einstein B coefficient

$$B_{12} = 2\pi^2 |\mu_{12}|^2/(3\varepsilon_0 hc^2). \tag{3.9}$$

There is a similar expression to eqn 3.8 for induced emission, except that it involves the coefficient B_{21}. However, since the same transition moment governs both up and down transitions, i.e. $\mu_{12} = \mu_{21}$, it follows that

$$B_{12} = B_{21} = B. \tag{3.10}$$

We see then that, when a sample is irradiated with light of the appropriate frequency, those molecules which are in the upper level are encouraged to move to the lower level with the same probability as those in the lower level move to the upper.

Net absorption or emission

The probabilities given, for example, in eqns 3.6 and 3.8 refer to the likelihood of a single molecule making a transition between the two levels. If we have a collection of molecules, there is another factor which must be taken into account, namely the populations N_1 and N_2 of the two levels. Downward transitions produce emission of radiation with intensity given by

$$I_{\mathrm{em}} = N_2 h\nu_{12}[A + B\rho(\nu_{12})] \tag{3.11}$$

while the upward transitions lead to absorption, with intensity

$$I_{\mathrm{abs}} = N_1 h\nu_{12} B \rho(\nu_{12}). \tag{3.12}$$

The overall effect is the difference between these two processes; whether it corresponds to net emission or absorption depends on whether N_2 is greater than N_1 or vice versa.

The most common situation is that of thermal equilibrium:

$$N_2/N_1 = \exp[-(E_2 - E_1)/kT]. \tag{3.13}$$

If N_2 is to be a significant fraction of N_1, the energy difference must be quite small (about 200 cm^{-1} at room temperature) in which case we can ignore the contribution made by spontaneous emission. On irradiating the sample, we would see net *absorption* proportional to the population difference $(N_1 - N_2)$ of the two levels.

It is also possible to arrange for there to be a non-equilibrium distribution between the two levels with N_2 greater than N_1. This is called a *population inversion*. If such an assembly is irradiated with light of the appropriate frequency, induced emission is observed and the initial radiation is amplified in intensity. This is the process which is exploited in lasers.

3.3 Absorption coefficients

For a *linear* absorption process, involving weak electromagnetic fields, the change in intensity dI of radiation passing through a distance dx of a sample is given by

$$dI = -I\alpha c dx \qquad (3.14)$$

where I is the intensity, c is the molecular (or molar) concentration, and α is known as the absorption coefficient. If we integrate over the total absorption path l, we obtain the familiar Beer–Lambert law:

$$I = I_0 e^{-\alpha l c}. \qquad (3.15)$$

It is the absorption coefficient α which contains the information on molecular energy levels and transition probabilities. For a molecular system with discrete quantum transitions, the absorption coefficient has a contribution from each of these transitions. Explicitly, α can be written

$$\alpha(\nu) = (h\nu/4\pi)\sum_{k,i} B_{ki}(N_i - N_k)g(\nu) \qquad (3.16)$$

where $g(\nu)$ is a normalized lineshape factor. The precise form of the lineshape function need not concern us here. It is sufficient to register that the spectroscopic line has a finite width in practice. It is the goal of high resolution spectroscopy to make this width as small as possible!

3.4 Spectroscopic selection rules and intensities

From what we have seen above, the essential factors for the description of the intensity of a spectroscopic absorption line are

$$\text{Intensity} \quad \propto \quad (N'' - N') < \Psi'|\mu_Z|\Psi'' >^2 \qquad (3.17)$$

where we have used $'$ and $''$ to refer to the upper and lower state in the transition and have assumed that the oscillating electric field in the light wave is polarized in the Z direction in the laboratory (so that it interacts with the Z component of the oscillating dipole moment). The all-important quantity in this discussion is therefore the transition moment. Its magnitude governs the intensity of the transition and its symmetry properties give the selection rules. (More correctly, its symmetry properties identify those transitions which are rigorously forbidden.)

The selection rules are expressed in terms of quantum numbers whose physical significance derives from the Born–Oppenheimer separation, discussed in Section 2.3. If we take

$$\Psi_{tot} = \Psi_{el}\Psi_{vib}\Psi_{rot} \tag{3.18}$$

and assume that the wavefunctions are real, the transition moment becomes

$$< \Psi'|\mu_Z|\Psi'' > = \int\int\int\Psi'_{el}\Psi'_{vib}\Psi'_{rot}\mu_Z\Psi''_{rot}\Psi''_{vib}\Psi''_{el}d\tau_{rot}d\tau_{vib}d\tau_{el} \tag{3.19}$$

where $d\tau_{rot}$ is the volume element for the rotational coordinates and so on. At this stage, the coordinates are expressed in an axis system of fixed orientation with its origin at the molecular centre of mass. We now transform to the molecule-fixed axis system (x, y, z) which rotates with the molecule, so that

$$\mu_Z = \sum_{\alpha=x,y,z}\lambda_{Z\alpha}\mu_\alpha. \tag{3.20}$$

In this equation, the quantity $\lambda_{Z\alpha}$ is a transformation coefficient between the two coordinate systems. It is known as a *direction cosine* and is a function in general of the three Euler angles which define the orientation of the molecule-fixed axis system, i.e. it is a function of the rotational coordinates. On the other hand, the dipole moment component μ_α is a function of electronic and nuclear coordinates measured in the molecule-fixed axis system. It is a function therefore of electronic and vibrational cordinates. The transformation to the molecule-fixed axis system allows the transition moment to be factorized

$$\text{trans moment} = \sum_\alpha\int\Psi'_{rot}\lambda_{Z\alpha}\Psi''_{rot}d\tau_{rot}\int\int\Psi'_{el}\Psi'_{vib}\mu_\alpha\Psi''_{vib}\Psi''_{el}d\tau_{vib}d\tau_{el}. \tag{3.21}$$

On the right hand side of eqn 3.21, the first factor depends on the rotational quantum numbers and gives the rotational selection rules. The second factor in this equation gives the vibrational and electronic selection rules. For example, for the *pure rotational spectrum*,

$$\Psi'_{el} = \Psi''_{el} \quad \text{and} \quad \Psi'_{vib} = \Psi''_{vib}$$

and the second integral in eqn 3.21 is just the permanent electric dipole moment of the molecule. Symmetry tells us therefore that the observation of a rotational transition requires the molecule to have a non-zero electric dipole moment (or at least a non-zero component). The first integral gives the rotational selection rules

$$\Delta J = 0, \pm 1 \tag{3.22a}$$

$$\Delta\Lambda \text{ (or } \Delta K) = 0 \quad \text{or} \quad \pm 1. \tag{3.22b}$$

If we turn our attention to the second, vibronic integral, we must accept that it cannot be factorized rigorously into a vibrational and an electronic part. Instead, we expand the dipole moment μ_α as a power series in the vibrational coordinate Q

$$\mu_\alpha = \mu_\alpha^e + \sum_i(\partial\mu_\alpha/\partial Q_i)_e Q_i + \frac{1}{2}\sum_{i,j}(\partial^2\mu_\alpha/\partial Q_i\partial Q_j)_e Q_iQ_j + \cdots\cdots \tag{3.23}$$

For a *vibrational* transition, the contribution from the first term on the right hand side of eqn 3.23 vanishes,

$$\int\int\Psi''_{el}\Psi'_{vib}\mu_\alpha^e\Psi''_{vib}\Psi''_{el}d\tau_{vib}d\tau_{el} = \int\Psi''_{el}\mu_\alpha^e\Psi''_{el}d\tau_{el}\int\Psi'_{vib}\Psi''_{vib}d\tau_{vib} = 0 \tag{3.24}$$

because the wavefunctions for two different vibrational levels are orthogonal. The leading contribution to the vibrational transition moment therefore comes from the second term which can be factorized into an electronic and a vibrational part

$$\int\Psi''_{el}(\partial\mu_\alpha/\partial Q)_e\Psi''_{el}d\tau_{el}\int\Psi'_{vib}Q\Psi''_{vib}d\tau_{vib} \neq 0. \tag{3.25}$$

The first factor tells us that, for a vibrational transition to be allowed, the dipole moment must change as the vibration is executed. The second factor gives us the vibrational selection rules, $\Delta v = \pm 1$ for a simple harmonic oscillator. The rotational selection rules for such a transition are given by the direction cosine factor in eqn 3.21, as discussed above.

Finally, we turn our attention to electronic transitions. For an *allowed electronic transition*, there must be a non-zero component of the dipole moment which oscillates at the transition frequency, i.e.

$$\int \Psi'_{el} \mu^e_\alpha \Psi''_{el} d\tau_{el} \neq 0. \tag{3.26}$$

This requires the direct product $\Gamma(\Psi_{el}') \times \Gamma(\mu_\alpha) \times \Gamma(\Psi_{el}'')$ to contain the totally symmetric representation of the molecular point group (Σ^+ for a diatomic molecule). If this condition is satisfied, we can again factorize electronic and vibrational parts

$$\text{trans moment} \propto \int \Psi'_{el} \mu^e_\alpha \Psi''_{el} d\tau_{el} \int \Psi'_{vib} \Psi''_{vib} d\tau_{vib}. \tag{3.27}$$

The first factor on the right hand side is the electronic transition moment and the second is the vibrational overlap factor, $S_{v'v''}$. The square of this latter integral is known as the Franck–Condon factor; it governs the relative intensities of different vibrational bands in an electronic spectrum. We shall discuss the part it plays in more details in section 7.5. As before, the associated rotational selection rules and relative intensities come from the direction cosine matrix elements in eqn 3.21.

The overlap integral $S_{v'v''}$ gives the vibrational selection rule. For a totally symmetric vibration, as is the case for a diatomic molecule, any change in the quantum number v_s is allowed. The relative intensities of the different bands are proportional to the Franck–Condon factor. For a non-symmetric vibration, the selection rule is

$$\Delta v_a = 0, 2, 4, 6... \tag{3.28}$$

because an odd change in v_a changes the symmetry of the vibrational state.

What happens when all three components of the electronic transition moment (eqn 3.26) are zero by symmetry? Despite the fact that the electronic transition is forbidden in this case, it is still possible to observe transitions between these electronic states provided that the vibronic transition moment

$$\int \int \Psi'_{el} \Psi'_{vib} \mu^e_\alpha \Psi''_{vib} \Psi''_{el} d\tau_{vib} d\tau_{el} \neq 0. \tag{3.29}$$

The transition is said to be *vibronically allowed*. It is the second term in the dipole moment expansion (eqn 3.23) which is primarily responsible for the transition intensity, i.e. we require

$$\int \Psi'_{el} (\partial \mu_\alpha / \partial Q)_e \Psi''_{el} d\tau_{el} \neq 0. \tag{3.30}$$

In this case, the vibrational selection rules (and relative intensities) are governed by the integral $\int \Psi'_{vib} Q \Psi''_{vib} d\tau_{vib}$ and, in consequence, the vibrational selection rule for a non-totally symmetric vibration is

$$\Delta v_a = 1, 3, 5... \tag{3.31}$$

Inversion and the parity selection rule

There is one more important selection rule for spectroscopic transitions, namely that involving the *parity* of the molecular eigenstate. Parity is another symmetry label which describes the behaviour of the wavefunction under laboratory-fixed inversion operator, E^*:

$$E^*f(X, Y, Z) \quad = \quad f(-X, -Y, -Z). \tag{3.32}$$

Because space is isotropic in the absence of external fields, a molecular wavefunction is transformed by inversion into either itself or the negative of itself:

$$E^*\Psi \quad = \quad \pm \Psi. \tag{3.33}$$

If the upper sign is obeyed, the state is said to have a *positive* or *even* parity; if on the other hand, the lower sign choice is obeyed, the state has *negative* or *odd* parity. Note that E^* is a symmetry operation for all molecules because it reflects a property of three-dimensional space. It should not be confused with i, the inversion operator for centro-symmetric molecules which acts in the molecule-fixed axis system:

$$i f(x, y, z) \quad = \quad f(-x, -y, -z). \tag{3.34}$$

Now it is easy to show that the electric dipole operator μ is an odd parity operator:

$$E^*\mu \ = \ E^*\sum_i q_i r_i \ = \ \sum_i q_i(-r_i) \ = \ -\mu. \tag{3.35}$$

In this equation, the summation is carried out over all charged particles, each of which carries a charge q_i. For a transition to be allowed, the transition moment $< \Psi' \mid \mu_Z \mid \Psi''>$ in eqn 3.17 must be non-zero which requires the integrand to be symmetric with respect to E^*. Since μ_Z is antisymmetric, the product of Ψ' and Ψ'' must be antisymmetric also. Therefore, the parity selection rule for electric dipole transitions is

$$+ \leftrightarrow - \text{ allowed}, \ + \rightarrow + \text{ and } - \leftrightarrow - \text{ forbidden}. \tag{3.36}$$

It is worth noting that the parity selection rule is different for magnetic dipole transitions because the magnetic moment is an even parity operator. The reason for this difference is that the magnetic dipole moment is associated with the angular motion of the charges in the molecule. Now the orbital angular momentum l of a single particle is

$$l = r \times p \ . \tag{3.37}$$

Thus, though r and p are individually antisymmetric with respect to E^*, their (vector) product is symmetric and so consequently is the magnetic moment **m**. Thus, for magnetic dipole transitions,

$$+ \leftrightarrow + \text{ and } - \leftrightarrow - \text{ are allowed}, \ + \leftrightarrow - \text{ forbidden}. \tag{3.38}$$

In passing, the selection rule for the g and u symmetry labels can also be deduced. These symmetry labels are only appropriate for a symmetric (homonuclear diatomic) molecule, of course. A vibronic state is said to be gerade (g) or ungerade (u) if

$$i\Psi_{\text{ev}}(q_{\text{el}}, q_{\text{vib}}) \quad = \quad \pm \Psi_{\text{el}}(q_{\text{ev}}, q_{\text{vib}}). \tag{3.40}$$

where i is defined in eqn 3.34. The arguments follow exactly the same lines as those for parity except that they are based on eqn 3.34 rather than 3.32. In consequence, electric dipole transitions for a centro-symmetric molecule are governed by the selection rule $g \leftrightarrow u$ whereas magnetic dipole transitions are allowed for $g \leftrightarrow g$ or $u \leftrightarrow u$.

4 Rotational spectroscopy

4.1 Introduction

In a gas phase sample, the molecules are in continual motion, appropriate to its thermal energy. Part of this is stored in the form of rotational energy, that is the kinetic energy associated with the tumbling motion of the molecules relative to an observer in the laboratory. For a single, isolated molecule, this rotational motion is quantized in accordance with quantum theory as described in Chapter 2. The transitions between these individual rotational levels give rise to the *rotational spectrum* of the molecule in question. The quantum states of a single molecule remain well defined as long as it is in isolation. If the molecule collides with another species, it is likely to change its rotational state and the definition of the spectrum is in danger of being destroyed. Provided the time between collisions is long compared with the inverse of the transition frequency (about 10^{-10} s), the rotational spectrum will remain sharply defined. This condition can be achieved in gases at low pressure (less than 10^{-4} atm). As the gas pressure is raised, the spectral lines become broader. In a high pressure gas or a liquid, the collision frequency is so high, and the spectral lines are consequently so broad, that individual transitions can no longer be seen. Under these conditions, the structural information contained in the transition frequencies is lost. In the solid state, the orientation of individual molecules is usually fixed in space, and energy cannot be stored in the form of rotational motion. Consequently, rotational spectroscopy is carried out on gaseous samples at low pressure (ideally sufficiently low that the spectral linewidth is limited by Doppler broadening rather than pressure broadening).

4.2 The rotational Hamiltonian: eigenvalues and eigenvectors

Classically, we can think of a diatomic molecule as two point masses, m_1 and m_2, situated at the ends of a massless, rigid rod of length r. This body rotates in space about its centre of mass with a kinetic energy equal to

$$E_{rot} = \tfrac{1}{2} I \omega^2 \qquad (4.1)$$

where I is the moment of inertia of the rigid body and ω is its angular velocity (measured relative to an axis system fixed in the laboratory). At any given instant in time, the orientation of a diatomic molecule can be defined by specifying the values of two angles, θ and ϕ, with respect to the laboratory-fixed axes. For a general (i.e. non-linear) body, three angles are required, θ, ϕ and χ. These are known as the Eulerian angles and are the *rotational coordinates*. The angular velocity ω is related to the partial derivatives of these coordinates with respect to time. The moment of inertia I is given by

$$I = m_1(z_1 - \bar{z})^2 + m_2(z_2 - \bar{z})^2 \qquad (4.2)$$

where the molecule-fixed z axis lies along the internuclear axis and $\bar{z}$ is the z-coordinate of the molecular centre of mass in this axis system. Usually the molecule-fixed axis system has its origin at the centre of mass and so $\bar{z}$ is zero. It is easy to show from eqn 4.2 that

$$I = \mu r^2 \tag{4.3}$$

where μ is known as the reduced mass of the molecule,

$$\mu = m_1 m_2 / (m_1 + m_2). \tag{4.4}$$

The reduced mass is a quantity which is introduced whenever motion in space is referred to the system's centre of mass. In the case of the rotational motion of a diatomic molecule, eqn 4.3 tells us that the rotational motion of the two masses, m_1 and m_2, about their centre of mass is exactly equivalent to the angular motion of a single particle of mass μ at a distance r from its axis of rotation. Thus a two-body problem is reduced to a (much simpler) one-body problem.

In the transformation from classical to quantum mechanics, we know how to transform position vectors r to their operator representatives but not velocity vectors v or angular velocity vectors ω. Rather, we know how to transform momenta, either linear (p) or angular (P) as discussed in section 2.2. We therefore rewrite the classical energy expression in eqn 4.1 as

$$E_{rot} = \tfrac{1}{2} P^2 / I \tag{4.5}$$

where P is the angular momentum:

$$P = I\omega. \tag{4.6}$$

The transformation to quantum mechanical operators can now be performed. The classical angular momentum P in eqn 4.5 is replaced by an operator $\hbar J$ where $\hbar = h/2\pi$ is the basic unit of angular momentum for quantum systems and J is a dimensionless operator. The explicit form of J is a moderately complicated function of partial differential operators, $\partial/\partial\theta$ and $\partial/\partial\phi$. With this replacement, the energy expression is also transformed to an operator, the rotational Hamiltonian H_{rot}, where

$$H_{rot} = (\hbar^2/2I)J^2. \tag{4.7}$$

In order to establish which energy states are allowed by quantum mechanics, we need to determine the eigenvalues and eigenfunctions of the rotational Hamiltonian, H_{rot}. The possible solutions follow directly from standard results of angular momentum theory (which is not covered in this book), namely that the operator J^2 satisfies the eigenvalue equation:

$$J^2 \Psi_{JM} = J(J+1)\Psi_{JM} \tag{4.8}$$

where J is the rotational angular quantum number and takes integral values only, $J = 0, 1, 2, 3$, etc. and Ψ_{JM} is the corresponding eigenfunction. The physical significance of eqn 4.8 is that only states with angular momentum of magnitude $\hbar [J(J+1)]^{1/2}$ are allowed by quantum mechanics. We shall discuss the form of these eigenfunctions shortly. Let us confine our attention to the allowed energy levels (eigenvalues) of H_{rot} for the moment:

$$E_{rot} = (\hbar^2/2I)\, J(J+1) \quad \text{where } J = 0, 1, 2, .. \tag{4.9}$$

This expression tells us that the rotational energy levels increase quadratically with J, in other words the spacings between successive levels get larger as J increases. The first few levels are plotted in Fig. 4.1 to show this behaviour.

We see from eqn 4.9 that the rotational energy levels depend *inversely* on the moment of inertia I, i.e. the larger the moment of inertia, the more closely spaced are the energy levels. The coefficient of $J(J+1)$ in the energy level expression, $\hbar^2/2I$, is the quantity which governs this spacing and is known as the *rotational constant*, B. Equation 4.9 is in energy units (Joules) and so if B is expressed in these units, it is defined by

$$B/\text{J} = \hbar^2/2I \tag{4.10}$$

Fig. 4.1 The first five rotational levels of a diatomic molecule.

where I is equal to μr^2. However, it is much more common to use either frequency units (Hz) or wavenumber units (cm^{-1}) in practice, in which case

$$\text{either} \qquad B/\text{Hz} = \hbar^2/2hI \tag{4.11}$$

$$\text{or} \qquad B/\text{cm}^{-1} = \hbar^2/2hcI. \tag{4.12}$$

For rotational spectroscopy, it is usual to measure the transition energy as a frequency. In frequency units, therefore, the rotational energy levels are given by

$$E_{\text{rot}}/h = BJ(J+1) \quad \text{where } J = 0, 1, 2, ... \tag{4.13}$$

4.3 Rotational wavefunctions, form and symmetry

We have introduced the rotational wavefunctions Ψ_{JM} in eqn 4.8. It turns out that these functions are a well known set of special functions, the *spherical harmonics,* with which the reader will probably already be familiar. They are functions of the two rotational coordinates, θ and ϕ, and depend on two quantum numbers, J and M. We have introduced J, the total angular momentum quantum number in the previous section. The other quantity, M, is known as the component (or sometimes the magnetic) quantum number. It is defined by the eigenvalue equation

$$J_Z \Psi_{JM} = M_J \Psi_{JM} \tag{4.14}$$

and M takes all possible (integral) values between $-J$ and $+J$, that is $(2J+1)$ separate values all together. Equation 4.14 shows that M_J measures the component of the total rotational angular momentum along the laboratory-fixed Z axis. Field-free space is isotropic and the choice of this axis is completely arbitrary. If an external field is applied to a molecule, the orientation of the field usually suggests a natural choice for the laboratory-fixed axis system. Equations 4.8 and 4.14 are statements of the fact that both the magnitude and orientation of a vector quantity like rotational angular momentum are quantized on the molecular scale. We see from eqn 4.13 that the rotational eigenvalues depend on J but are independent of the quantum number M, in the absence of external electric or magnetic fields. In other words, each rotational level J is $(2J+1)$-fold degenerate. This property is important when we come to consider the population of a level at a particular temperature or transitions between these levels.

Table 4.1   Rotational wavefunctions[a]

$J = 0$	$\Psi_{JM} = Y_{0,0} = \sqrt{1/4\pi}$
$J = 1$	$\Psi_{JM} = Y_{1,0} = \sqrt{3/4\pi}\cos\theta$
	$\Psi_{JM} = Y_{1,\pm1} = \mp\sqrt{3/8\pi}\sin\theta\,e^{\pm i\phi}$
$J = 2$	$\Psi_{JM} = Y_{2,0} = \sqrt{5/16\pi}(3\cos^2\theta - 1)$
	$\Psi_{JM} = Y_{2,\pm1} = \mp\sqrt{15/8\pi}\sin\theta\cos\theta\,e^{\pm i\phi}$
	$\Psi_{JM} = Y_{2,\pm2} = \sqrt{15/32\pi}\sin^2\theta\,e^{\pm 2i\phi}$

[a] Normalization: $\int_0^{2\pi}\int_0^{\pi}\Psi_{JM}{}^*\Psi_{JM}\sin\theta\,d\theta\,d\phi = 1$.

The explicit forms of the wavefunctions for the first three rotational levels are given in Table 4.1. In order to appreciate these fully, the reader has only to realize that they are exactly the same as the angular part of the eigenfunctions for the H atom. Thus, the wavefunction for the lowest level, $J = 0$, is simply the familiar *s*-orbital, the three wavefunctions for $J = 1$ with $M = 1, 0$, and -1 are the three *p*-orbitals, and so on. For the lowest levels at least, a good qualitative picture of each wavefunction already exists.

In order to define the selection rules of transitions between rotational states, one must first appreciate the symmetry properties of the wavefunctions. There are two useful symmetry operations, (i) rotation through an infinitesimally small angle about an axis through the centre of mass and (ii) inversion through the origin of the chosen laboratory-fixed axis system. These operations reflect the isotropic nature of three-dimensional space. From the point of view of the measurements we can make on it, a quantum system is unaffected by changing its orientation or turning it inside out. Although we have introduced the rotational quantum number J to define the energy levels in eqn 4.9, it is in fact a symmetry label (together with M) and gives the transformation properties of the wavefunction under the operations of the rotation group.

We have already met the inversion operator E^* in Section 3.4. Its effect on the rotational wavefunctions in eqn 4.8 is

$$E^*\Psi_{JM} = (-1)^J\Psi_{JM}. \qquad (4.15)$$

Each rotational state therefore carries a parity symmetry label, + for even J levels and − for odd J levels. These are shown in Fig. 4.1.

4.4   Selection rules and the rotational spectrum

The selection rules which govern transitions between the rotational levels of a molecule in a given vibrational level of a particular electronic state have been discussed in the previous chapter, Section 3.4. For electric dipole transitions of a diatomic molecule, the rules are:

$$\Delta J = 0, \pm1 \qquad (4.16a)$$

$$\Delta\Lambda = 0 \qquad (4.16b)$$

$$\text{parity}\quad + \leftrightarrow -. \qquad (4.16c)$$

The quantum number Λ is the component of both $\mathbf{L}$, the orbital angular momentum, and $\mathbf{J}$ along the internuclear axis. The molecule must also possess a permanent electric dipole moment; in other words, it must be heteronuclear.

Table 4.2   Rotational constants for some typical diatomic molecules

molecule	B_0/cm^{-1}	B_0/GHz
$H_2{}^a$	59.3219	1778.43
HD	44.662	1338.93
HF	20.5567	616.274
HCl	10.4398	312.978
CO	1.92253	57.6360
$N_2{}^a$	1.98958	59.6461
CS	0.817085	24.4956
CuCl	0.177741	5.23855

[a] Homonuclear molecule, has zero dipole moment and therefore no rotational spectrum.

Only the selection rule $\Delta J = +1$ is relevant for the rotational spectrum of a diatomic molecule in a closed shell, $^1\Sigma^+$ state; the parity selection rule is subsumed by it. Applying this selection rule to the energy level scheme in eqn 4.13, the allowed transitions occur at frequencies v where

$$hv = B[(J+1)(J+2) - J(J+1)] = 2B(J+1). \qquad (4.17)$$

In other words, the spectrum consists of a set of equally spaced lines starting at a frequency of $2B/h$ and separated by $2B/h$. Some typical B-values for diatomic molecules are given in Table 4.2. From this, it can be seen that rotational transitions occur in the range from 10 GHz (the microwave region) through 100 GHz (the sub-millimetre wave region) up to 1 THz (the far-infrared region) depending on the molecule being studied. It is therefore strictly not correct to refer to rotational spectroscopy generically as microwave spectroscopy.

As we saw in Section 3.4, the intensity of an absorption line is proportional to the product $v (N'' - N') < \Psi'|\mu_Z|\Psi'' >^2$. For a gas phase sample in thermal equilibrium at a temperature T, the population difference factor becomes

$$(N'' - N') = N''(1 - e^{-hv/kT}). \qquad (4.18)$$

Since $hv \ll kT$ in the region covered by rotational spectroscopy, the population of the upper M-state involved in the transition is not negligible compared with that of the lower state and the absorption intensity is a measure of the population difference between the two M-states. Expanding the exponential term as a power series, we see that this population difference increases linearly with the transition frequency v

$$(N'' - N') = N'' (1 - 1 + hv/kT - ....). \qquad (4.19)$$

In this expression, N'' is the population of an individual M state of the lower level in the transition. When expressed in terms of the population of the $J = 0$ level, N_0, this becomes

$$N'' = N_0 e^{-BJ(J+1)/kT}. \qquad (4.20)$$

We turn now to the transition moment factor. For a $J + 1 \leftarrow J$ transition of a diatomic molecule, we require the direction cosine matrix element

$$\int\int\int\Psi^*_{J'M'}\lambda_{Zz}\Psi_{J''M''}\sin\theta d\theta d\phi=\delta_{M'M''}[\{(J+1)^2-M^2\}/\{(2J+1)(2J+3)\}]^{1/2}.$$
$$(4.21)$$

Thus the intensity of the transition between individual states $M \leftarrow M$ is proportional to $\mu_e^2 \{(J+1)^2 - M^2\}/\{(2J+1)(2J+3)\}$, where μ_e is the permanent electric dipole moment. In the absence of external fields, all the individual transitions labelled with different M values occur at the same frequency. The total absorption intensity is therefore given by the sum over all possible M values ($M = -J$ to $+J$). Using

$$\sum_M 1 = (2J+1) \tag{4.22a}$$
$$\sum_M M^2 = \tfrac{1}{3}J(J+1)(2J+1) \tag{4.22b}$$

we obtain the final result

$$\text{rel intensity} \propto \nu\,(h\nu/kT)N_0 e^{-BJ(J+1)/kT}\tfrac{1}{3}(J+1)\mu_e^2. \tag{4.23}$$

Since, from eqn 4.17, the transition frequency ν is proportional to $(J+1)$, the J-dependence goes as $(J+1)^3 e^{-BJ(J+1)/kT}$. This function is plotted against J in Fig. 4.2. As the frequency increases, the lines get stronger up to a maximum beyond which they decrease as the Boltzmann exponential factor starts to dominate. It is sometimes asserted that the intensity goes as the population of the lower J-level in the transition,

$$\text{population} = (2J+1)e^{-BJ(J+1)/kT}. \tag{4.24}$$

This is clearly not correct. The proper expression has the effect of allowing the rotational spectrum to be followed out to higher J values than suggested by the simple consideration of population. A simulation of the rotational spectrum of CO at 300 K is given in Fig. 4.3.

4.5   Centrifugal distortion

Moderately careful measurements of the rotational spectrum of a diatomic molecule show that the lines are not exactly spaced at $2B/h$ but get slightly closer as J increases. This is a manifestation of *centrifugal distortion*. The bond between the two nuclei is not infinitely stiff and, as the molecule rotates faster, the masses move apart, thereby increasing the moment of inertia and so reducing the rotational constant.

This phenomenon can be described by a simple, classical treatment. Let us assume that, as the bond is stretched, the restoring force is described by Hooke's Law:

$$\mathbf{F} = -k(\mathbf{r} - \mathbf{r_e}) \tag{4.25}$$

where k is the force constant. This amounts to saying that the stretching motion is that of a simple, harmonic oscillator. Our picture of the rotating molecule is that of two masses at the ends of a stiff spring, tumbling over in space (rather than at the ends of a rigid rod). If we next refer the motion to the centre of mass, as in Section 4.2, we again reduce the problem to a one-body problem, namely the rotation of a single mass μ about a fixed point restrained by a spring of force constant k. As this system rotates at an angular velocity ω, the bond (or spring) stretches to a length $\mathbf{r}$ until the restoring force is large enough to provide the requisite acceleration towards the origin to keep the

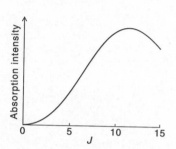

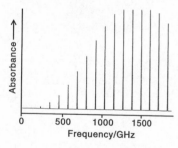

Fig. 4.2   The variation of the relative absorption intensities of rotational transitions in a diatomic molecule with J.

Fig. 4.3   The rotational spectrum of CO at room temperature. Note the strong dependence of the intensities on frequency (or J).

orbit stable. In other words, the stretching of the bond provides the centripetal force:

$$k(\mathbf{r} - \mathbf{r_e}) = \mu r \omega^2. \tag{4.26}$$

Solving this equation for the instantaneous bond length r, we obtain

$$r = r_e/[1 - \mu \omega^2/k] \tag{4.27}$$

so that the faster the molecule rotates, the longer the bond length (as expected).

The stretching of the bond makes two separate contributions to the energy of rotation, one in the form of kinetic energy and the other as potential energy. To evaluate the kinetic contribution, we write

$$1/I = 1/\mu r^2 = (1 - \mu \omega^2/k)^2/\mu r_e^2$$
$$= 1/\mu r_e^2 - 2\mu P^2/(I^3 k) + ... \tag{4.28}$$

where P is the rotational angular momentum. Therefore the kinetic energy is

$$\tfrac{1}{2} P^2/I = \tfrac{1}{2} P^2/I_e - \mu P^4/I_e^3 k + ... \tag{4.29}$$

where we have neglected the distinction between I and I_e. The first term on the right hand side is the rigid rotor kinetic energy; the second term is a contribution to the centrifugal distortion correction.

The potential energy contribution is given by

$$V_{cd} = \tfrac{1}{2} k(r - r_e)^2 = \tfrac{1}{2} k(\mu r \omega^2/k)^2 , \tag{4.30}$$

from eqn 4.26. Substituting P/I for ω, we obtain

$$V_{cd} = \tfrac{1}{2} \mu P^4/I_e^3 k . \tag{4.31}$$

The total centrifugal distortion correction is the sum of the contributions in eqns 4.29 and 4.31

$$E_{cd} = -\tfrac{1}{2} \mu P^4/I_e^3 k \tag{4.32}$$
$$\Rightarrow -\tfrac{1}{2} [h^4/I_e^3 (k/\mu)] J^2 (J + 1)^2 \tag{4.33}$$

making the transfer to quantum mechanical form. Taking the centrifugal distortion effects into account, the rotational kinetic energy of a diatomic molecule is written

$$E_{rot} = B J(J + 1) - D J^2 (J + 1)^2 \tag{4.34}$$

where D is known as the centrifugal distortion constant and is equal to

$$D = \tfrac{1}{2} \hbar^4/I_e^3 (k/\mu) \tag{4.35}$$

in energy units. We shall see in the next chapter that the harmonic vibrational frequency of a diatomic molecule is related to the force constant k of the bond by

$$\nu_{harm}/Hz = \tfrac{1}{2\pi} \sqrt{(k/\mu)}. \tag{4.36}$$

Substitution of this result in eqn 4.35 gives

$$D/\text{Hz} = 4B^3/\nu_{harm}{}^2. \tag{4.37}$$

The same result can be obtained by a straightforward quantum mechanical calculation, using second order perturbation theory as we shall see in Chapter 5. Since the ratio B/ν_{harm} is of the order of 10^{-3}, the centrifugal distortion correction is a very small part of the rotational kinetic energy for all reasonable values of J. Equation 4.37 shows that the value of the centrifugal distortion parameter D can be calculated reliably from a knowledge of B and ν_{harm}. Conversely, a measurement of D from the rotational spectrum provides an estimate of the harmonic vibrational frequency.

Incorporating the effects of centrifugal distortion, the frequency of a transition $J + 1 \leftarrow J$ in the rotational spectrum is calculated to be

$$\nu = 2B(J+1) - 4D(J+1)^3. \tag{4.38}$$

Values for B and D can be determined most conveniently in practice by dividing eqn 4.38 by $(J+1)$:

$$\nu/(J+1) = 2B - 4D(J+1)^2. \tag{4.39}$$

Thus a plot of $\nu/(J+1)$ against $(J+1)^2$ gives a straight line with intercept $2B$ and slope $-4D$.

4.6   Determination of bond lengths; vibrational averaging

Rotational spectroscopy provides a very precise measurement of the bond angles and bond lengths which constitute the molecular geometry. (Only the bond length is relevant for a diatomic molecule, of course.) More explicitly, the rotational spectrum yields a value for B, the rotational constant, which can be related to the moment of inertia by eqn 4.11, for example. The inertial properties are then related to the molecular geometry. Thus, in the rotational spectrum of a diatomic molecule, we do not measure r, the internuclear bond length, directly; rather, we determine a value for r^{-2}.

In order to interpret the measured bond length correctly, we need to understand how it is derived in a bit more detail. We have seen in the previous section that, as the molecule rotates in space, its bond length fluctuates comparatively rapidly as a result of vibrational motion. What is measured in the rotational spectrum therefore is the moment of inertia averaged over this vibrational motion, $<\Psi_v|(\mu r^2)^{-1}|\Psi_v>$. The effect of the vibrational averaging depends on the form of the wavefunction and in particular on the value of the vibrational quantum number v. As a result, the rotational constant shows a weak vibrational dependence which can be modelled by the expression

$$B_v = B_e - \alpha_e(v + \tfrac{1}{2}) + \dots \tag{4.40}$$

The vibrational coefficient α_e has two distinct contributions, which are called harmonic and anharmonic; we shall discuss this in more detail in the next chapter. The parameter is typically a few percent of B_v in magnitude and has a positive sign for a stretching vibration (as in a diatomic molecule); in other words, B_v gets slightly smaller as v increases. We know from eqn 4.10 that B_e is proportional to μ^{-1}; α_e on the other hand is proportional to $\mu^{-3/2}$.

The simplest experiment which can be carried out in practice is a measurement of B_0 (and D_0) for a diatomic molecule in its zero point level. The value for the bond length which is obtained from B_0 is referred to as r_0, it being recognized that this is a value averaged over the $v = 0$ wavefunction. If a more accurate (and meaningful) bond length is required, the rotational spectrum must be recorded for the molecule in several vibrational levels $v = 0, 1, 2, ...$ The B_0 values so determined allow accurate extrapolation back to the B_e value by eqn 4.40. This is known as the equilibrium rotational constant and provides a value for r_e, the equilibrium bond length (that is, the separation of the two nuclei at the minimum of the vibrational potential curve, see Section 5.1). Careful measurement allows this bond length to be measured very accurately, of the order of 1 part in 10^8. At this level, one can see signs of breakdown of the Born–Oppenheimer separation.

4.7 The Stark effect: molecules in electric fields

When the spectrum of a molecule is recorded in the presence of an external electric field, the lines split up into a number of subcomponents, the exact number depending on the J values involved. This phenomenon is known as the Stark effect, after the German physicist who first discovered the effect for the hydrogen atom in 1913.

In order to understand this behaviour, we need first to see how an electric field affects the energy levels of a diatomic molecule. From a classical viewpoint, we know that, in the presence of an electric field $\mathbf{E}$, the energy of a molecule changes with orientation, strictly with the orientation of its dipole moment, μ:

$$W_{\text{Stark}} = -\mu \cdot \mathbf{E}. \tag{4.41}$$

The minus sign represents the fact that the energy is lowest when the dipole moment and the electric field are parallel and highest when they are anti-parallel. We note that this interaction is exactly the same as that which forms the basis of electric dipole transitions given in eqn 3.2. It differs only in the precise form of the field. For the Stark effect, we apply a static, time-independent field whereas, to drive electric dipole transitions, we must apply a field which oscillates at the frequency appropriate to the transition. Despite this difference, the quantum mechanical treatment of the effect is identical to that given in Chapter 3.

Let us assume that we apply a uniform electric field of magnitude E_0 along the laboratory-fixed Z axis. We convert the general classical expression eqn 4.41 to quantum mechanical operator form and then transfer the electric dipole operator so that it can act within the molecule-fixed axis system

$$H_{\text{Stark}} = -\mu \cdot \mathbf{E} = -E_0 \sum_\alpha \lambda_{Z\alpha}(\theta, \phi, \chi)\mu_\alpha \tag{4.42}$$

where, as before, $\lambda_{Z\alpha}(\theta, \phi, \chi)$ is the direction cosine between the laboratory-fixed Z axis and the molecule-fixed α ($= x, y$ or z) axis. It is a function of the three Euler angles (i.e. of the rotational coordinates). Invoking the Born–Oppenheimer separation, we can factorize the expectation value of H_{Stark} into a vibronic part and a rotational part. For a diatomic molecule with cylindrical symmetry, the vibronic part $<\Psi_{\text{vib}}\Psi_{\text{el}} |\mu_\alpha| \Psi_{\text{el}}\Psi_{\text{vib}}>$ is zero for $\alpha = x$ or y; the only non-zero component of μ lies along the internuclear or z axis. For $\alpha = z$, the direction cosine operator connects rotational states with $\Delta J = 0, \pm 1$, $\Delta\Lambda = 0$ and $\Delta M = 0, \pm 1$. Furthermore, the parity labels of the two overall states must be opposite, $+ \leftrightarrow -$.

We see from these selection rules that the rotational wavefunctions Ψ_{JM} are not eigenfunctions of H_{Stark} since states with J differing by 1 and of opposite parity are mixed together by the interaction. When the field is large, the mixing becomes so pronounced that it is no longer possible to assign a particular value for J or a particular parity label to a given eigenstate. The quantum numbers J and parity are no longer *good quantum numbers*. The physical reason for this result is as follows: The presence of the electric field destroys the isotropy of free space and rotational angular momentum is no longer conserved. The electric field imposes its own, lower (cylindrical) symmetry on space. As a result, the quantum number M, the component of J along the electric field direction, remains a good quantum number. The physical effect of applying an external electric field is to cause the molecule to rotate in a cockeyed manner about the field direction rather than having its rotational angular momentum vector point in a fixed direction in space.

In practice, there are two types of Stark effect, the *first order* and the *second order* effect. The first order effect requires that the rotational energy levels have either vibrational or electronic degeneracy. Only the latter is possible for diatomic molecules and, even then, it is very unusual. The one familiar example is nitric oxide, NO, which has a $^2\Pi$ electronic ground state; each of its rotational levels is two-fold degenerate, the two components having opposite parity. The $\Delta J = 0$ matrix elements of the Stark Hamiltonian connect these two components. When an electric field E_0 is applied, the energy levels split according to the formula

$$E_{\text{Stark}} = \mu_z E_0 \Omega M / J(J+1) \qquad (4.43)$$

where μ_z is the dipole moment and Ω is the component of the total angular momentum (rotational plus electronic) along the internuclear axis. This splitting is shown as a function of E_0 in Fig. 4.4. Note that each level remains two-fold degenerate in the presence of the field and it is not possible by experimental observation to assign a sign to the M value to the states involved. Equation 4.43 shows that the splitting of the different M components of the rotational level and also of the spectral line depends linearly on the electric field and on M. Since this is a first order effect, large splittings can be produced by quite modest fields (a few hundred V cm^{-1}).

Most molecules studied by rotational spectroscopy show second order Stark effects. These molecules are in closed-shell electronic states and each rotational level has only its $(2J+1)$ degeneracy, arising from the different M states; each of these M states is non-degenerate and has the same parity (+ for even J and – for odd J). Thus there is no effect from the $\Delta J = 0$ terms of the Stark Hamiltonian. The splittings which are observed in practice involve the $\Delta J = \pm 1$ matrix elements. A level J is mixed with levels $J+1$ and $J-1$ by the electric field to produce the second order Stark effect; this can be described well for reasonable fields by second order perturbation theory. The result of this calculation is a correction to the energy as follows

$$W_{\text{Stark}}^{(2)} = [\mu_z^2 E_0^2 / \{2hBJ(J+1)\}][\{J(J+1) - 3M^2\}/\{(2J-1)(2J+3)\}]$$

$$(4.44)$$

where B is in frequency units. The shifts of the M states are much smaller than those caused by the first order effect. The shifts are proportional to the *squares* of the major quantities involved, μ_z, E_0 and M. As with the the first order effect, there is a residual two-fold degeneracy of M states which is not lifted by the electric field, i.e. for the second order effect, we cannot distinguish the

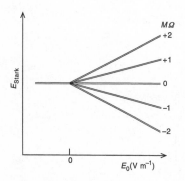

$M\Omega$
+2
+1
0
−1
−2

Fig. 4.4 The splitting of the M components of the $J = 2$ level of a molecule in a $^1\Pi$ state. The energy depends linearly on the electric field because of the twofold degeneracy (at zero field) due to Λ-doubling.

levels with $+M$ and $-M$. As a result of this, it is not possible to determine the *sign* of the dipole moment from a Stark measurement, that is, the direction in which μ points in the molecule. Usually, this is obvious from simple considerations of electronegativity—but not always. The behaviour of the levels under the second order Stark effect is shown in Fig. 4.5.

The Stark effect is exploited in molecular spectroscopy to measure the dipole moments of molecules, or more strictly their magnitudes for the reasons given above. Measurements made on rotational spectra give the most accurate and reliable values for this important molecular property.

The Stark effect also provides a valuable means of modulation of the molecular transition frequencies in an experimental spectrometer. If the static field applied in eqn 4.41 is replaced by a slowly oscillating one (slow compared with the oscillating electromagnetic field), the frequencies of the various Stark components will move up and down in frequency at twice the modulation frequency. It is possible in practice to use this modulation to increase the signal-to-noise ratio and hence the sensitivity of the apparatus.

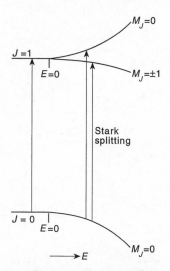

Fig. 4.5 The second-order Stark effect on the first rotational transition in a polar diatomic molecule.

5 Vibrational spectroscopy

5.1 Introduction

As we move up the energy scale, the next hierarchy of energy levels which we encounter after rotation is that associated with the vibrational motion of molecules. Vibrational motion is a periodic, concerted displacement of the nuclei (more correctly, of the atoms) in a molecule which leaves the centre of mass unaltered in laboratory space (we recall that the motion of the molecular centre of mass is described as translation). The appropriate linear combination of the displacements of each nucleus from its equilibrium position is called the *vibrational coordinate*, q and is used to describe a particular vibrational motion. For example, in the linear molecule CO_2, the symmetric stretching coordinate q_1 involves the displacement of the two O atoms away from (or towards) each other. There is no contribution from the C atom to the vibrational coordinate because it remains fixed at the centre of mass (Fig. 5.1). A general, polyatomic molecule has several, distinct vibrational modes but, for the special case of the diatomic molecule, there is only one. The motion in this case is the stretching of the bond between the two atoms. Many alternative definitions of the vibrational coordinate for a diatomic molecule are possible but the simplest is

$$q = r - r_e \tag{5.1}$$

where r and r_e are the instantaneous and equilibrium bond lengths, respectively.

The energy associated with vibrational motion can now be expressed in terms of this vibrational coordinate. There are two distinct contributions to the energy, the kinetic part which arises from the motion of the nuclei and the potential part, which comes from the compression or expansion of the bond from its equilibrium value. As the nuclei move back and forth, the energy (which is constant for an isolated molecule) is transferred between the kinetic and potential forms. When the nuclei are at their equilibrium separation, the energy is solely kinetic whereas, when the nuclei are stationary at the turning points of the motion, the energy is wholly potential. We see from this that the potential energy varies as a function of the vibrational coordinate q. The typical form of this function for a diatomic molecule is shown in Fig. 5.2; it is often referred to as the *potential energy curve*.

It is worthwhile taking the time to understand the physical meaning of a potential energy curve before we consider the detailed description of vibrational motion. We recall from Chapter 2 that the electrons in a molecule move so much faster than the nuclei that, to a first approximation, we can solve the equations of motion for the electrons with the nuclei fixed at some particular separation r. The result of such a calculation gives us the energy of the rigid molecule in a given electronic state. If we repeat the calculation with a slightly different value of r, we obtain a different energy. Repetition of this process for successive values of the parameter r allows the potential energy

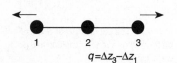

Fig. 5.1 The vibrational coordinate q_1 which describes the symmetric stretching vibration of the CO_2 molecule. Note that there is no contribution from the position of the C atom to this coordinate.

curve in Fig. 5.2 to be traced out. This curve therefore tells us how the kinetic energy of the electrons plus the electrostatic energy of the charged particles in the molecule vary as a function of r. Of course, in a real molecule, the nuclei move as well (in the form of vibrations) but on a much longer time scale than the period of electronic motion. This nuclear motion takes place within the confines of the potential energy curve and can be described in a separate quantum mechanical calculation. We shall consider this calculation later in the chapter.

The form of the potential energy curve given in Fig. 5.2 shows that, at short internuclear distances, the molecular energy increases rapidly because the charged particles in the molecule experience strong repulsive forces when they are close together. At the opposite extreme, if a bond is stretched enough, it will break. In the limit of large values of r, the potential energy of the system becomes constant, being simply the sum of the energy of two independent atoms. This is the process of dissociation and the difference in energy between the dissociation limit and the minimum of the curve is known as the *dissociation energy*, D_e. Although an exact mathematical expression for the potential energy curve of a general diatomic molecule is not known, a simple and convenient one which reproduces the qualitative shape of the curve shown in Fig. 5.2 has been suggested by P. M. Morse, namely:

$$U(q) = D_e[1 - e^{-\beta q}]^2 \tag{5.2}$$

where D_e is the dissociation energy, q is the distortion of the bond from its equilibrium value, and β is a constant for any given molecule (it defines the narrowness of curvature of the potential well at its bottom). For any given molecule, the Morse potential energy curve will be in only approximate agreement with the true one. Nevertheless, it does provide a useful description of the general distribution of the vibrational energy levels and the properties of the vibrational states.

The vibrational motion of a typical diatomic molecule causes only small displacements of the bond from its equilibrium length. In vibrational spectroscopy therefore, we shall be particularly interested in the form of the potential energy curve near its minimum (small values of q). This suggests that a suitable expression for the potential energy can be obtained by a Maclaurin series expansion about $q = 0$:

$$U(q) = U_0 + (\mathrm{d}U/\mathrm{d}q)_0 q + \tfrac{1}{2}(\mathrm{d}^2 U/\mathrm{d}q^2)_0 q^2 + \dots \tag{5.3}$$

We are free to choose $U = 0$ at $r = r_e$ so that the leading term on the right hand side of eqn 5.3 vanishes. Furthermore, at $q = 0$, the potential energy is a minimum and therefore $(\mathrm{d}U/\mathrm{d}q)_0$ must be zero also. The leading term in the expansion is therefore quadratic

$$U(q) \approx \tfrac{1}{2}(\mathrm{d}^2 U/\mathrm{d}q^2)_0 q^2. \tag{5.4}$$

We expect this approximation, in which only the leading term is retained, to be valid for values of r near the equilibrium position. The form of the potential curve in eqn 5.4 corresponds to the assumption that the restoring force generated as the molecular bond is stretched follows Hooke's law. For a true Hooke's law spring, $\mathrm{d}^2 U/\mathrm{d}q^2$ is equal to the force constant k for all values of q, and U is a parabola. For a chemical bond, the potential energy curve only approximates to a parabola near the minimum. This is exemplified in Fig. 5.3

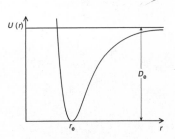

Fig. 5.2 The vibrational potential energy function, plotted with respect to a bond length r.

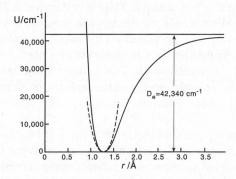

Fig. 5.3   The Morse potential energy function for the diatomic molecule HCl. The dotted curve shows the parabola which matches this curve at its minimum and defines its harmonic properties.

which shows the Morse function which corresponds to the r_e and the D_e values for HCl and the parabola which fits this curve near the minimum. When the potential energy curve can be approximated by a parabola, the vibrational motion is that of a *harmonic oscillator*. We discuss the quantum mechanical solution to the harmonic oscillator problem in more detail in the next few sections.

5.2   The simple harmonic oscillator: reduction to one-body form

We shall first describe the classical vibrational motion of a diatomic molecule in some detail so that we can make the most favourable choice of coordinates. The two-body problem is depicted in Fig. 5.4. The two atoms are approximated by point masses m_1 and m_2 which are at distances r_1 and r_2 from the centre of mass at any given instant. The displacements of each atom from its equilibrium position are given by

$$q_1 = r_1 - r_{1e} \tag{5.5a}$$

$$\text{and} \qquad q_2 = r_2 - r_{2e}. \tag{5.5b}$$

The classical expression for the vibrational energy is

$$W_{\text{class}} = \text{k.e.} + \text{p.e.} \tag{5.6}$$

The kinetic energy part is given by

$$\text{k.e.} = \tfrac{1}{2} m_1 v_1^2 + \tfrac{1}{2} m_2 v_2^2 \tag{5.7}$$

where v_1 and v_2 are the instantaneous velocities of particles 1 and 2. The condition for the centre of mass requires that

$$m_1 r_1 = m_2 r_2$$

$$\text{i.e.} \qquad m_1 q_1 = m_2 q_2. \tag{5.8}$$

Now the total displacement $(q_1 + q_2)$ is simply what we have referred to as the vibrational coordinate q before. Hence

$$q_1 = m_2 q / (m_1 + m_2) \tag{5.9a}$$

$$q_2 = m_1 q / (m_1 + m_2). \tag{5.9b}$$

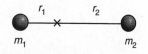

Fig. 5.4   The diatomic molecule as a harmonic oscillator, represented as a two-body problem. The centre of mass is marked with an X.

Therefore

$$m_1 v_1^2 = m_1 [m_2/(m_1 + m_2)]^2 \dot{q}^2 \qquad (5.10)$$

with a similar expression for $m_2 v_2^2$. Using these results, the k.e. term in eqn 5.6 simplifies to

$$\tfrac{1}{2} m_1 v_1^2 + \tfrac{1}{2} m_2 v_2^2 = \tfrac{1}{2} [m_1 m_2/(m_1 + m_2)] \dot{q}^2 \equiv \tfrac{1}{2} \mu \dot{q}^2. \qquad (5.11)$$

Here we have introduced the reduced mass μ again, defined earlier in eqn 4.4; its involvement always tells us that we have referred our description to the centre of mass of the system. Because we know how to transform momenta to quantum mechanical operators but not velocities, we shall rewrite eqn 5.11 as

$$\text{k.e.} = \tfrac{1}{2} p^2/\mu \qquad (5.12)$$

where $p = \mu \dot{q}$, the vibrational momentum.

We now turn our attention to the potential energy term. In the harmonic approximation, this is

$$\text{p.e.} = \tfrac{1}{2} k_1 q_1^2 + \tfrac{1}{2} k_2 q_2^2. \qquad (5.13)$$

Now, if k is the force constant for stretching the whole bond, that for each section is increased inversely in proportion to its length:

$$k_1 = k r/r_1 = k q/q_1. \qquad (5.14a)$$

Similarly

$$k_2 = k q/q_2. \qquad (5.14b)$$

Therefore

$$\text{p.e.} = \tfrac{1}{2} k q(q_1 + q_2) = \tfrac{1}{2} k q^2. \qquad (5.15)$$

Collecting the k.e. and p.e. contributions, we can write

$$W_{\text{vib}} = \tfrac{1}{2} p^2/\mu + \tfrac{1}{2} k q^2. \qquad (5.16)$$

In other words, we have reduced the original two-body problem to a one-body description of vibration which is much easier to solve from a quantum mechanical point of view. The reduced, one-body problem can be envisaged as the vibrational motion of a particle of mass μ against a fixed point, under the restraining influence of a spring of length r with a force constant k. This is shown in Fig. 5.5.

Fig. 5.5 The one-body harmonic oscillator which corresponds to the two-body oscillator in Fig. 5.4. A particle of mass μ vibrates against a fixed point under the influence of a spring of length r and force constant k.

5.3 Eigenvalues and eigenfunctions of the simple harmonic oscillator

Solution of the Schrödinger equation

We can now write down the Hamiltonian operator which represents the motion of a simple harmonic oscillator, using the prescription introduced in Chapter 2:

$$q \rightarrow \mathbf{q} \qquad (2.1)$$

$$p \rightarrow \mathbf{p} = -i\hbar \partial/\partial q \qquad (2.2)$$

and

$$W_{\text{class}} \rightarrow H_{\text{vib}}.$$

Hence

$$H_{\text{vib}} = -(\hbar^2/2\mu) \mathrm{d}^2/\mathrm{d}q^2 + \tfrac{1}{2} k q^2. \qquad (5.17)$$

This is the Hamiltonian for the simple harmonic oscillator. If we write a general eigenfunction for this operator as $\psi_v(q)$, the Schrödinger equation becomes

$$-(\hbar^2/2\mu)\mathrm{d}^2\Psi_v/\mathrm{d}q^2 + \tfrac{1}{2}kq^2\Psi_v = E_v\Psi_v. \tag{5.18}$$

In other words, we need to find the solutions of a second-order, ordinary differential equation. Although this is a relatively simple problem to solve, it is somewhat lengthy. The two essential steps are:

1. Express $\psi_v(q)$ in the trial form $\Psi_v = \exp(-\alpha q^2/2)\,H(q)$ where $\alpha^2 = k\mu/\hbar^2$. This form ensures that the wavefunction shows the correct behaviour as $q \to \pm\infty$, namely that $\psi_v(q)$ tends to zero.

2. Solve the 'standard' differential equation for the factor $H(q)$

$$\mathrm{d}^2H/\mathrm{d}\xi^2 - 2\xi\mathrm{d}H/\mathrm{d}\xi + (\lambda/\alpha - 1)H = 0 \tag{5.19}$$

where $\lambda = 2\mu E_v/\hbar^2$ and ξ is a dimensionless coordinate related to q by

$$\xi = q/r_{\mathrm{e}}. \tag{5.20}$$

This is a commonly occuring differential equation and its solutions form a class of 'special functions', so much so that they are given their own name. They are in fact a set of polynomials and are known as the *Hermite polynomials*, $H_v(\xi)$. (It is slightly disturbing that very similar symbols are used for both the vibrational Hamiltonian and the Hermite polynomials. Fortunately, the distinction between the two is usually clear in context.)

Eigenvalues of the simple harmonic oscillator

Because the motion of the reduced particle of mass μ is confined by the boundaries of the potential energy curve, the energy levels associated with its motion are quantized. The eigenvalues which satisfy eqn 5.18 are

$$E_v = (v + \tfrac{1}{2})hv \tag{5.21}$$

where v is the vibrational quantum number which can take only integral values 0, 1, 2, 3 ... and v is the harmonic frequency in cycles/sec (Hz)

$$v = \tfrac{1}{2\pi}(k/\mu)^{1/2}. \tag{5.22}$$

The unit of relative energy most commonly used for vibrational spectroscopy is the wavenumber unit, cm^{-1}. The vibrational wavenumber, the inverse of the associated wavelength, is distinguished from the frequency by writing it as ω_{e}:

$$E_v = (v + \tfrac{1}{2})hc\omega_{\mathrm{e}}. \tag{5.23}$$

The pattern of energy levels described by eqn 5.23 is shown in Fig. 5.6.

Three important properties of the pattern of energy levels associated with the simple harmonic oscillator should be appreciated.

First, we note that the energy levels are equally separated and form a uniform ladder with the spacing

$$E_{v+1} - E_v = hc\omega_{\mathrm{e}} \tag{5.24}$$

independent of v. This very simple property follows from the pleasingly

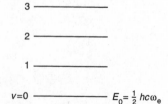

Fig. 5.6 The first few energy levels of a harmonic oscillator, labelled by the vibrational quantum number v. The lowest level is not at zero but possesses zero-point energy due to the residual motion.

symmetric form of the classical energy in eqn 5.16 in which both the kinetic and potential energy terms are quadratic in form.

Second, we note from eqn 5.22 that the vibrational frequency is proportional to the square root of the force constant k and inversely proportional to the square root of the reduced mass μ. Thus, as the force constant increases (i.e. the bond of the molecule becomes stiffer), the separation between the vibrational levels–which is what we mean by the vibrational frequency, see eqn 5.24—also increases. As the reduced mass increases, the vibrational frequency decreases. In the limit as $k \rightarrow 0$ (or $\mu \rightarrow \infty$), the parabolic potential fails to confine the reduced particle and the energy is no longer quantized. The particle is then free and its energy, which is now translational, is continuous.

Third, we note from eqn 5.23 that the lowest energy of the harmonic oscillator with $v = 0$ is

$$E_0 = \tfrac{1}{2}hc\omega_e \tag{5.25}$$

i.e. it is not zero. For this reason, it is referred to as the *zero-point energy*. It is a manifestation of the Heisenberg Uncertainty Principle; we shall return to this point later.

Eigenfunctions of the simple harmonic oscillator

The general form of the eigenfunctions of the simple harmonic oscillator has been introduced earlier in this section. When expressed in its normalized form, it is

$$\Psi_v(q) = N_v H_v(y) \exp(-y^2/2) \tag{5.26}$$

where y is a dimensionless coordinate, defined by

$$y = 2\pi(\mu v/h)^{1/2}q \tag{5.27}$$

and the normalization constant $N_v = (v!\pi^{1/2}/2^v)^{1/2}$ ensures that

$$\int \Psi_v(q)^2 \mathrm{d}q = 1. \tag{5.28}$$

We note that the wavefunction in eqn 5.26 is real. $H_v(y)$ are the Hermite polynomials. For $v = 0$, $H_0(y) = 1$; consequently, the wavefunction is simply proportional to the Gaussian function $\exp(-y^2/2)$. For $v = 1$, $H_1(y) = 2y$; the wavefunction is now the same Gaussian function multiplied by $2y$. The first few Hermite polynomials are given in Table 5.1. It can be seen that the Hermite polynomials get progressively more complicated as v increases; H_v is a polynomial of degree v. The corresponding wavefunctions are shown in Fig. 5.7. The point where the wavefunction crosses through zero is called a *node*. The wavefunction for level v has v nodes.

Table 5.1 Hermite polynomials

v	$H_v(y)$
0	1
1	$2y$
2	$4y^2 - 2$
3	$8y^3 - 12y$
4	$16y^4 - 48y^2 + 12$

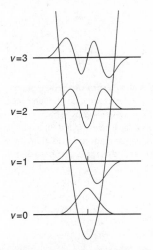

Fig. 5.7 The form of the eigenfunctions for the first few levels of a harmonic oscillator. The point where the wavefunction crosses through zero is known as a node.

5.4   The quantum harmonic oscillator: non-classical behaviour

One of the difficulties in understanding the behaviour of quantum mechanical systems is that it can differ fundamentally from the expectations based on classical, Newtonian physics. An appreciation of this non-classical behaviour is often crucial to understanding experimental measurements on these systems. The simple harmonic oscillator demonstrates this non-classical behaviour in three separate ways:

First, remembering that the probability distribution of the oscillator in a level v is given by the square of the wavefunction $\psi_v{}^2$, we see from Fig. 5.7 that there is a small but finite probability that the oscillating particle can be outside the classically accessible region. (The classical system cannot move outside the confines of the parabolic potential curve in Fig. 5.7.) This is an example of the phenomenon known as *quantum mechanical tunnelling*. The extent of tunnelling depends on the steepness of the potential wall (we recall, for example, that for the particle in a box problem, with infinite vertical walls, there is no tunnelling). Thus, tunnelling for a simple harmonic oscillator is most pronounced at low v. As the quantum number increases, the wall becomes steeper and tunnelling is less evident. This behaviour is in line with the Correspondence Principle.

The second, dramatic example of non-classical behaviour concerns the probability distribution of the simple harmonic oscillator within the classical region. For example, in the level $v = 0$, the probability distribution is $\psi_0{}^2$. Since the square of a Gaussian is also a Gaussian, the probability distribution is proportional to $\exp(-y^2)$. This distribution is shown in Fig. 5.8(a). It can be seen that the particle is most likely to be found at $q = 0$. This is exactly the opposite of classical expectation. For a classical harmonic oscillator, a pendulum for example, the particle is most likely to be found at its turning points (on the parabola) because here its velocity is zero; it is *least* likely to be found at $q = 0$. The classical probability distribution also has the shape of a parabola; this is shown in Fig. 5.8(b). Mentally squaring the wavefunctions shown in Fig. 5.7, we can see that as v increases, the probability of the particle being at its turning points also increases. In the limit of large v, the behaviour goes over to the classical motion, yet another example of the Correspondence Principle.

We have already mentioned the third aspect of non-classical behaviour, namely the zero-point motion of a quantum oscillator. In the lowest possible level, with $v = 0$, the system is still in motion and therefore possesses energy. There is no way in which the oscillator can lose this energy; in other words, it cannot relax down to the bottom of the potential well. If it were able to do so, we would be in the position of knowing both its position ($q = 0$) and its momentum ($p = 0$) exactly. This conflicts with the Heisenberg uncertainty principle which states that the product of the uncertainty of the position δq and the momentum δp must be greater than a small but finite quantity

$$\delta p \delta q \geq \hbar/2 \tag{5.29}$$

i.e. it is not possible to define both position and momentum exactly. If the position is known exactly, the momentum becomes infinitely uncertain.

(a)

$P(q)$

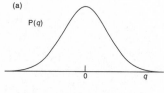

0 q

(b)

$P_{class}(q)$

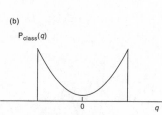

0 q

Fig. 5.8   The probability distribution function for a harmonic oscillator (a) for the $v = 0$ level where the maximum probability is at $q = 0$ and (b) the classical probability distribution, approached by the quantum system at large v.

5.5 Vibrational selection rules and spectrum of the simple harmonic oscillator

The selection rules governing a vibrational spectrum have already been discussed in Section 3.4. The transition moment for electric dipole transitions can be written in the molecule-fixed axis system as

$$\text{transition dipole moment} = \int \int \Psi'_{el} \Psi'_{vib} \mu_z \Psi''_{vib} \Psi''_{el} d\tau_{vib} d\tau_{el}. \qquad (5.30)$$

For a diatomic molecule, the dipole moment can be expanded as a Taylor series in the single vibrational coordinate q:

$$\mu = \mu_e + (\partial\mu/\partial q)_e q + \tfrac{1}{2}(\partial^2\mu/\partial q^2)_e q^2 + \dots. \qquad (5.31)$$

For the various terms on the right hand side of this equation to be non-zero, we require that the electronic expectation values of μ_e, $(\partial\mu/\partial q)_e$, and so on are non-zero. This is automatically satisfied if the molecule has a permanent dipole moment, i.e. the diatomic molecule is heteronuclear. (Strictly speaking, we see that all the electronic expectation values are zero for a homonuclear molecule. By default therefore, the molecule must be heteronuclear to have a vibrational spectrum.)

The leading term on the right hand side of eqn 5.31, μ_e, is independent of q and so has only diagonal elements in v. It therefore cannot bring about a change in v and does not contribute to the vibrational spectrum. The second term, linear in q, consequently makes the dominant contribution to the spectrum. For simple harmonic oscillator wavefunctions, this term has non-zero matrix elements with $\Delta v = 1$ only:

$$\langle v + 1|q|v \rangle = A[v + 1]^{1/2} \qquad (5.32)$$

$$\text{where } A = (\hbar/4\pi\mu v)^{1/2}. \qquad (5.33)$$

From this, the dominant vibrational selection rule is $\Delta v = 1$. Because the separation between the vibrational levels is usually much larger than kT, most of the molecules are in the $v = 0$ level at room temperature. Therefore the strongest band in the vibrational spectrum, called the *fundamental band*, is the $v = 1 \leftarrow 0$ transition. This transition occurs at the harmonic frequency ν or wavenumber ω_e.

Even at room temperature, there will be a few molecules in the $v = 1$ level. When illuminated with light of the correct frequency (ν for a harmonic oscillator), these molecules can be further excited to the level $v = 2$ through the interaction with the radiation. This transition is intrinsically twice as strong as the fundamental band because the intensity is proportional to $(v + 1)$. However, in practice the band is much weaker because of the low population of the $v = 1$ level. If this population is increased by making the sample warmer, the band will gain strength; for this reason, it is called a *hot band*. In the harmonic approximation, it occurs at the same wavenumber as the fundamental band. For an anharmonic oscillator, it occurs at a slightly different wavenumber, as we discuss below.

There is a third type of vibrational transition which can occur for a harmonic oscillator. It arises from the third term in the expansion of μ (eqn 5.31), that proportional to q^2. For q^2

$$\langle v + 2|q^2|v \rangle = A^2[(v + 2)(v + 1)]^{1/2}. \qquad (5.34)$$

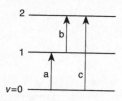

Fig. 5.9 The three possible types of vibrational transition for a harmonic oscillator. That labelled (a) is the fundamental band, (b) is a hot band, and (c) is an overtone.

The selection rule governing this transition is therefore $\Delta v = 2$. Consequently, we expect to see a weak transition $v = 2 \leftarrow 0$ at a wavenumber of $2\omega_e$ for a simple harmonic oscillator; it is weak because the second derivative of the dipole moment with respect to q is usually much smaller than the first derivative. This band is called the *overtone* or *harmonic band*, specifically the first overtone. The second, third, etc. overtones arise in a similar manner at $3\omega_e$, $4\omega_e$, ..., becoming progressively weaker.

The three types of vibrational transition which can occur for a diatomic molecule are summarized in Fig. 5.9. Values for the harmonic wavenumbers of some representative diatomic molecules are given in Table 5.2 together with their corresponding force constants. It can be seen that vibrational spectra are expected to occur within the range 300 to 4000 cm^{-1}.

Table 5.2 Harmonic wavenumbers for some diatomic molecules

Molecule	ω_e/cm^{-1}	k/N m^{-1}
HF	4138.32	966.189
HCl	2990.999	516.331
DCl	2145.154	516.331
CO	2169.814	1901.863
I_2[a]	214.502	172.013

[a] Homonuclear molecule, has zero dipole moment and therefore no vibrational transitions.

5.6 The anharmonic oscillator

We have already noted earlier in this chapter that the vibrational motion of a real diatomic molecule differs somewhat from that of a harmonic oscillator. The potential energy curve is not a parabola. Rather it is steeper on its inner wall reflecting the strong, repulsive forces as the two nuclei get very close. In addition, it falls away to a horizontal line at large values of r as the molecule dissociates. The general form of the potential energy curve for an anharmonic oscillator is shown in Fig. 5.2.

The energy levels of an oscillator vibrating within the confines of such an anharmonic potential are obtained from the Schrödinger equation of the same form as in eqn 5.18 but with the harmonic potential replaced by the anharmonic one. Although this is quite a simple exercise for a modern, electronic computer, the numerical procedure does not provide great physical insight. The general form of the anharmonic eigenfunctions can more easily be appreciated by adopting the Morse potential function given earlier. This casts the potential energy curve in a relatively simple, analytic form:

$$U(q) = D_e[1 - e^{-\beta q}]^2. \tag{5.2}$$

We recall that, as $q \to \infty$, $U \to D_e$, the disssociation energy. Also, for small displacements from the equilibrium position, the exponential term can be expanded to give

$$U(q) \approx D_e \beta^2 q^2. \tag{5.35}$$

In other words, the oscillator shows harmonic behaviour near r_e with a force constant k given by

$$k = 2D_e\beta^2. \tag{5.36}$$

Using this expression for the potential energy in the Schrödinger equation, exact solutions for the eigenvalues and eigenfunctions can be obtained. The eigenvalues of the so-called Morse oscillator are

$$G(v) = E_{vib}/hc = (v + \tfrac{1}{2})\omega_e - (v + \tfrac{1}{2})^2\omega_e x_e \tag{5.37}$$

where the parameter $\omega_e x_e$ is a positive quantity, known as the vibrational anharmonicity constant. The quantity x_e is dimensionless, typically of a magnitude of a few parts in 100. The form of these solutions is shown in Fig. 5.10. It can be seen that the vibrational levels get closer together as the dissociation limit is approached. The separation between adjacent vibrational levels $\Delta G(v)$ decreases *linearly* with v:

$$\Delta G(v) \equiv G(v + 1) - G(v) = \omega_e - 2(v + 1)\omega_e x_e. \tag{5.38}$$

When this spacing goes to zero, we move from a quantized region of energy levels to a continuum; this marks the onset of dissociation. Thus we see that, for a realistic potential energy function, there are only a finite number of vibrational energy levels below dissociation.

If the last level before dissociation is that with a quantum number v_L, setting $\Delta G(v)$ in eqn 5.38 to zero we obtain

$$(v_L + 1) = \omega_e/2\omega_e x_e. \tag{5.39}$$

This gives us the number of bound levels for a given potential energy function, because the first (lowest) level has $v = 0$. We can also make an estimate of the dissociation energy D_0 measured from the zero point level. For a Morse oscillator, this is

$$\begin{aligned} D_0 &= v_L\omega_e - v_L(v_L + 1)\omega_e x_e \\ &= \tfrac{1}{4}(\omega_e^2/\omega_e x_e) - \tfrac{1}{2}\omega_e. \end{aligned} \tag{5.40}$$

The dissociation energy D_e measured from the minimum of the potential energy curve is obtained from D_0 by adding the zero point energy:

$$D_e = \tfrac{1}{4}(\omega_e^2/\omega_e x_e). \tag{5.41}$$

(This is the exact relationship which is obtained from the solution of the Schrödinger equation using the Morse potential.)

We have established in the previous section that there are three types of vibrational transition expected for a diatomic molecule. Using the energy levels given in eqn 5.37, we see that the fundamental band occurs at a wavenumber of

$$\nu_{1-0} = \omega_e - 2\omega_e x_e. \tag{5.42}$$

The first hot band, $v = 2 \leftarrow 1$, occurs at

$$\nu_{2-1} = \omega_e - 4\omega_e x_e \tag{5.43}$$

i.e. at a wavenumber which lies $2\omega_e x_e$ to lower energy of the fundamental band. The first overtone band occurs at

$$\nu_{2-0} = 2\omega_e - 6\omega_e x_e; \tag{5.44}$$

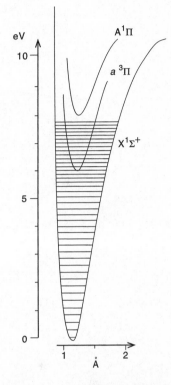

Fig. 5.10 The vibrational energy levels of an anharmonic oscillator, in this case the real example of CO. Note how the levels converge as one moves up the potential well.

this transition therefore does not occur at exactly twice the fundamental. It can be seen from these equations that measurement of either the hot or the overtone band along with the fundamental allows values for both ω_e and $\omega_e x_e$ to be determined.

The eigenfunctions of the Morse oscillator are of course different from those of the harmonic oscillator but can be expressed as linear combinations of them. We have seen in the previous section how the overtone bands can acquire intensity through the expansion of the electric dipole operator. The mixing of the simple harmonic wavefunctions by the potential energy terms for an anharmonic oscillator provides a second, independent mechanism to give this transition intensity. The two contributions are referred to as electrical and mechanical anharmonic contributions respectively.

The eigenvalues of the Morse oscillator are given exactly by eqn 5.37. For a more general anharmonic potential function, such as a converging power series expansion in q, the eigenvalues can be cast in a similar form except that it contains other contributions of higher power in $(v + \frac{1}{2})$:

$$G(v) = (v + \tfrac{1}{2})\omega_e - (v + \tfrac{1}{2})^2 \omega_e x_e + (v + \tfrac{1}{2})^3 \omega_e y_e + ... \qquad (5.45)$$

This suggests that the Morse oscillator expression will give only an approximate description of actual energy levels and is particularly likely to fail for large values of v, near the dissociation limit.

5.7   Energy levels of the vibrating rotator

So far, we have only considered the vibrational motion of a diatomic molecule. Provided that the molecule is not fixed in space (i.e. it is not in a solid state environment), it will be free to rotate also. However, the time scale for rotational motion is much longer than that for vibrational motion. The period for rotational motion is of the order of 10^{-11} s whereas that for vibrational motion is about 10^{-14} s. In other words, for each completed rotational cycle, the molecule vibrates back and forth about one thousand times. This difference in the time scales for the two motions is really the justification for the extension of the Born–Oppenheimer approximation to treat vibrational and rotational motions separately, as discussed in Chapter 2. We first solve the vibrational problem within the molecule-fixed axis system; this solution is independent of the molecular orientation. Once this has been done, we can average the vibrationally dependent variables over the appropriate vibrational wavefunction and so solve the rotational problem.

One approach to the determination of the vibration–rotation energy levels of a diatomic molecule is to adopt a zeroth order Hamiltonian which is the sum of the Hamiltonians for the anharmonic oscillator and for the rigid rotor, that is for a molecule with some fixed bond length, r_e:

$$H_{\text{vib-rot}}^{(0)} = H_{\text{vib}} + H_{\text{rot}}^{(0)} \qquad (5.46)$$

where
$$H_{\text{rot}}^{(0)} = (\hbar^2/2\mu r_e^2)J^2. \qquad (5.47)$$

The effects of vibrational motion on the rotational kinetic energy can then be added in by perturbation theory. These effects are small so long as the Born–Oppenheimer separation is valid (which it usually is).

The eigenvalues and eigenfunctions of the zeroth order Hamiltonian can be written down directly because the two terms in eqn 5.46 operate separately on independent variables, namely vibrational (q) and rotational (θ,ϕ) coordinates. Thus

$$E_{\text{vib-rot}}/hc = G(v) + B_e J(J+1) \qquad (5.48)$$

where $G(v)$ is given in general by the expression in eqn 5.45 and B_e is the equilibrium rotational constant (in cm^{-1}):

$$B_e = (\hbar^2/2hc\mu r_e^2). \qquad (5.49)$$

The expression 5.48 tells us that each vibrational level, as depicted in Fig. 5.11, has its own set of rotational energy levels, which are quadratically dependent on the quantum number J and possess a degeneracy ($2J + 1$), exactly as described for the $v = 0$ level in Chapter 4. In line with the Born–Oppenheimer separation, the spacing between adjacent rotational levels is approximately 10^3 times smaller than that between vibrational levels. The eigenfunction is a simple product of the vibrational and rotational eigenfunctions

$$\Psi^{(0)} = \Psi_{\text{vib}}(q)\chi_{\text{rot}}(\theta, \phi). \qquad (5.50)$$

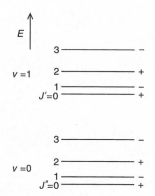

Fig. 5.11 The energy level pattern for a vibrating rotor. Each vibrational level has its own set of rotational levels.

We are now in a position to answer the question: What effect does the vibrational motion of a diatomic molecule have on its rotational energy levels? We have in fact already answered this question in Sections 4.5 and 4.6: the levels are subject to centrifugal distortion effects and the rotational constant varies smoothly with v. However, we can now describe these effects more clearly since we know more about vibrational motion.

The effects of vibration–rotation interactions can be incorporated in a relatively simple manner by the use of perturbation theory. The Hamiltonian is separated into two parts, a zeroth-order part $H^{(0)}$ which has known eigenvalues (and eigenfunctions) and the remainder which is called the perturbation V:

$$H_{\text{vib-rot}} = H_{\text{vib-rot}}^{(0)} + V. \qquad (5.51)$$

It is more straightforward to use the harmonic oscillator Hamiltonian for the zeroth-order vibrational operator since its eigenfunctions are easier to deal with mathematically. Thus the zeroth-order Hamiltonian is:

$$H_{\text{vib-rot}}^{(0)} = H_{\text{vib}}^{(0)} + H_{\text{rot}}^{(0)} \qquad (5.52)$$

where $H_{\text{vib}}^{(0)}$ is the harmonic oscillator Hamiltonian (eqn 5.17), and $H_{\text{rot}}^{(0)}$ is the rigid rotor Hamiltonian (eqn 5.47). The eigenfunctions of this zeroth order part are given in eqn 5.50; in other words, there is complete separation of vibrational and rotational coordinates at the zeroth-order level.

The perturbation term V sweeps up all the terms omitted from eqn 5.52 in the zeroth-order Hamiltonian; it can also be written as the sum of vibrational and rotational parts. The vibrational part V_{vib}, describes the anharmonic corrections to the vibrational potential which can be expressed as a power series expansion in q:

$$V_{\text{vib}} = g\,q^3 + h\,q^4 + \dots \qquad (5.53)$$

It is convenient to re-express this expansion in terms of the dimensionless anharmonic parameters a_1, a_2, a_3, etc. introduced by Dunham:

$$V_{vib} = \tfrac{1}{2}k\,q^2[a_1(q/r_e) + a_2(q/r_e)^2 + ...]. \tag{5.54}$$

The cubic parameter g is therefore given by

$$g = \tfrac{1}{2}ka_1/r_e = (hc\omega_e^2/4B_e)a_1. \tag{5.55}$$

We note that V_{vib} is independent of the rotational coordinates. Next we deal with the rotational perturbation term, V_{rot}. The full rotational Hamiltonian has exactly the same form as that given in eqn 5.48 except that the equilibrium bond length r_e is replaced by the instantaneous bond length r. Now we know from eqn 5.1 that

$$r = r_e + q$$
$$= r_e(1 + q/r_e). \tag{5.56}$$

Since the rotational kinetic energy operator involves r^{-2}, we express this quantity as a power series in q through eqn 5.56:

$$r^{-2} = r_e^{-2}[1 - 2(q/r_e) + 3(q/r_e)^2 - ...]. \tag{5.57}$$

Thus we can write the rotational Hamiltonian in terms of its zeroth-order part and a perturbation which is a function of the vibrational coordinate q:

$$H_{rot} = H_{rot}^{(0)} + V_{rot} \tag{5.58}$$

where $\qquad V_{rot} = -(\hbar^2/2\mu r_e^2)[2(q/r_e) - 3(q/r_e)^2 + ...]\boldsymbol{J}^2. \tag{5.59}$

We are now ready to apply perturbation theory to the description of the effects of vibration–rotation interactions. The centrifugal distortion correction to the rotational energy level formula

$$F(J) = BJ(J + 1) - DJ^2(J + 1)^2 \tag{5.60}$$

arises from the leading term in the rotational perturbation expansion, namely the one linear in q. We have seen in eqn 5.32 that such an operator has non-zero matrix elements in a simple harmonic oscillator basis set with $\Delta v = \pm 1$ only. Therefore this term in the rotational kinetic energy mixes the rotational levels of a particular level v with the levels immediately above $(v + 1)$ and below $(v - 1)$. The second order correction to the energy from this mixing is given by

$$E_v^{(2)} = \sum_{v' \neq v}\langle v|V_{rot}|v'\rangle\langle v'|V_{rot}|v\rangle/(E_v^{(0)} - E_{v'}^{(0)}) \tag{5.61}$$

where $E_v^{(0)}$ is the zeroth order energy of the harmonic oscillator level v. Substitution of the perturbation operator in eqn 5.59, using eqn 5.32 to evaluate the matrix element of the term linear in q, leads to a contribution to the energy which is quadratic in $J(J + 1)$. This has exactly the form of the centrifugal distortion correction in eqn 5.60 with the parameter D given by:

$$D = 4B_e^3/\omega_e^2. \tag{5.62}$$

This relationship, which has been derived here by quantum mechanics, is exactly the same as that in Chapter 4 (eqn 4.37), where it was obtained using classical arguments.

We now turn our attention to the vibrational dependence of the rotational constant B. As stated in Chapter 4, this is given by:

$$B_v = B_e - \alpha_e(v + \tfrac{1}{2}) + ... \tag{5.63}$$

The quantum mechanical calculation shows that there are two separate contributions to the parameter α_e. The first of these is the *harmonic* contribution; it arises as a first order contribution to the energy from the term quadratic in q in V_{rot} (eqn 5.59):

$$E^{(1)} = \langle v|V_{rot}|v \rangle. \tag{5.64}$$

Using the result:

$$\langle v|q^2|v \rangle = 2A^2(v + \tfrac{1}{2}) \tag{5.65}$$

derived from eqn 5.32, we obtain:

$$E^{(1)}/hc = 6(B_e^2/\omega_e)(v + \tfrac{1}{2})J(J + 1). \tag{5.66}$$

This correction to the rotational constant is caused by vibrational averaging and occurs even for a harmonic oscillator; it has the effect of making the constant *larger* as v increases. We can see physically how it arises from Fig. 5.12, which shows the variation of the inverse of the moment of inertia I with r. For a harmonic oscillator, the classical turning points of the vibrational motion are equidistant from r_e; put another way, the square of the wavefunction Ψ_v is symmetrical about r_e. In the evaluation of the expectation value of I^{-1} over the wavefunction Ψ_v, the positive contribution on the smaller r side of r_e outweighs the negative contribution on the larger side. Thus, $(I^{-1})_v > I_e^{-1}$ and so $B_v > B_e$.

The second part of α_e is derived from the vibrational perturbation to the potential function and so is called the anharmonic contribution. Its calculation by perturbation theory involves the cross term in second order between the term linear in q in V_{rot} and the cubic term in V_{vib}. As with the centrifugal distortion term, the second order mixing arises from the adjacent vibrational levels, $(v + 1)$ and $(v - 1)$. The end result is:

$$E^{(2)}/hc = 6a_1(B_e^2/\omega_e)(v + \tfrac{1}{2})J(J + 1). \tag{5.67}$$

Since the anharmonic parameter a_1 is always negative, so also is this contribution to the rotational kinetic energy. The effective value for r therefore increases with v from these interactions. This is because the anharmonic potential leans outwards as r increases and so weights larger values of r more heavily in the quantum mechanical averaging process. Combining the harmonic and anharmonic contributions, we obtain

$$\alpha_e = -6(B_e^2/\omega_e)(a_1 + 1). \tag{5.68}$$

The value of a_1 is usually in the range between -2 and -4 and so the anharmonic (positive) contribution to α_e outweighs the harmonic (negative) contribution. We thus arrive at the familiar result that the rotational constant B_v of a diatomic molecule *decreases* with increasing vibrational excitation.

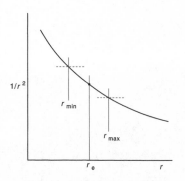

Fig. 5.12 The variation of the quantity r^{-2} with r. In taking the average over the harmonic oscillator wavefunction, the portion for r below r_e outweighs that for r above r_e so that the average of r^{-2} is greater than r_e^{-2}.

5.8 The vibration–rotation spectrum

Selection rules

Transitions between the levels of a molecule which is both rotating and vibrating are fortunately governed by selection rules (fortunately because this makes the spectrum simpler). For electric dipole transitions, these are:

$$\Delta J = \quad 0, \pm 1 \tag{5.69a}$$

$$\text{parity} \quad + \leftrightarrow - \tag{5.69b}$$

and

$$\Delta v = \quad 1, 2, \dots \tag{5.69c}$$

The selection rules were introduced earlier in Section 3.4 and the vibrational selection rules have been discussed in more detail in Section 5.5. The implication of these rules is that both the vibrational and rotational state are changed as a result of the transition. For each vibrational transition governed by the selection rule eqn 5.69c, several rotational transitions can occur from different, populated levels. Since the separation between rotational energy levels is so much smaller than that between vibrational levels, the rotational transitions are comparatively closely grouped about a common origin. Such a collection of lines is called a *vibration–rotation band*.

The selection rule on J shows that three types of transition are expected. Since these occur in different regions of the spectrum, they are often distinguished by the labels P, Q, and R (corresponding to $\Delta J = -1$, 0, and $+1$, respectively). For the great majority of diatomic molecules, namely those in closed-shell electronic states, the Q-lines are forbidden. This is because, although allowed by the dipole selection rule, they would violate the parity selection rule for electric dipole transitions. Rotational levels with the same J value in different vibrational levels always have the same parity (+ for even J and − for odd J). Therefore a transition which obeys $\Delta J = 0$ connects states with same parity and so is not allowed. This point can perhaps be appreciated more easily by reference to Fig. 5.11.

The infrared spectra of almost all diatomic molecules therefore show only P and R branches. An example of an exception to this statement is nitric oxide, NO, which is an open-shell molecule with a $^2\Pi$ ground state. This molecule has all three branches in its infrared spectrum, including the Q branch. The reason for this different behaviour is that each rotational level is two-fold degenerate, one component having + parity and the other having − parity (a phenomenon known as lambda-type doubling). Therefore, for any given rotational level, there is always a state of the appropriate parity for the transition to terminate on and so be allowed.

Intensities

As discussed in Chapter 3, the relative intensity of an absorption line in the vibration–rotation spectrum is proportional to the product $v(N'' - N') < \Psi'|\mu_Z|\Psi'' >^2$. For rotational spectroscopy in the microwave region, the population of the two levels is comparable and the intensity is markedly affected by the *population difference*. Vibration–rotation transitions occur in the infrared region. This energy separation corresponds to a very large difference in the population of the upper and lower levels (at room temperature) and, for all but the very lowest vibrational frequencies, the population difference $(N'' - N')$ can be reliably replaced by N''. We have seen in Section 3.4 that, for vibration – rotation transitions, the transition moment can be factorized into a rotational part and a vibrational part. The latter is essentially constant for a given vibrational band. The rotational factor can be calculated as described in Section 4.4. The matrix element of the direction cosine operator between individual M states of the upper and lower rotational levels is first calculated. The square of this quantity is then summed over all possible values of M' and M'' to give the rotational factor, known as the *line strength*, $S_{J'J}$. For the two possible transitions, this quantity is directly proportional to J. Specifically,

$$\text{P, } \Delta J = -1: \qquad S_{J'J} = \tfrac{1}{3}J \qquad (5.70a)$$

$$\text{R, } \Delta J = +1: \qquad S_{J'J} = \tfrac{1}{3}(J+1). \qquad (5.70b)$$

Therefore the J dependence of the relative intensity is given by

$$I_{\text{rel}} \propto S_{J'J}\exp[-hcB''J(J+1)/kT]. \qquad (5.71)$$

The two factors in this expression are plotted as a function of J in Fig. 5.13(a); their product is given in Fig. 5.13(b). It is seen that in both P and R branches, the lines get stronger as J increases, pass through a maximum, and then decrease as the exponential Boltzmann factor starts to bite.

The strongest line in the P and R branches can be identified from eqn 5.71. If we take the derivative of I_{rel} for the R branch with respect to J and equate the result to zero, we can determine the J value J_{max} which corresponds to the strongest line:

$$J_{\text{max}} = \tfrac{1}{2}[(2kT/hcB) + \tfrac{1}{4}]^{1/2} - \tfrac{3}{4}. \qquad (5.72)$$

This value is close to, but not the same as, the value of J for the most populated level in the lower state:

$$J_{\text{pop max}} = \tfrac{1}{2}(2kT/hcB)^{1/2} - \tfrac{1}{2}. \qquad (5.73)$$

The value for J_{max} which can be identified experimentally is often used to estimate the temperature of the sample.

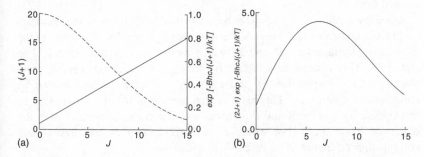

Fig. 5.13 The relative intensities of lines in the R branch of a vibration–rotation band of a diatomic molecule. Part (a) shows the dependences of the linestrength and the Boltzmann population factor separately, as a function of a J and part (b) shows their product (which is the relative intensity).

Appearance of a vibration–rotation band

The wavenumber ν of a transition in a vibration–rotation band is just the difference between the upper and lower state energy levels, i.e.

$$\nu = G(\nu') - G(\nu'') + F(J') - F(J''). \qquad (5.74)$$

The difference of the two vibrational terms gives the band origin ν_0. Thus for the fundamental band, for example,

$$\nu_0 = \omega_e - 2\omega_e x_e + \dots \tag{5.75}$$

from eqn 5.42. The wavenumbers of individual lines in the two branches can be calculated by substituting for the rotational terms in eqn 5.74. Neglecting for the moment the effects of centrifugal distortion, we obtain

$$\nu_P = \nu_0 - (B' + B'')\,J + (B' - B'')\,J^2 \tag{5.76}$$

and

$$\nu_R = \nu_0 + 2B' + (3B' - B'')\,J + (B' - B'')\,J^2$$

$$= \nu_0 + (B' + B'')\,(J+1) + (B' - B'')\,(J+1)^2. \tag{5.77}$$

Both these equations can be represented by the single expression

$$\nu = \nu_0 + (B' + B'')m + (B' - B'')m^2 \tag{5.78}$$

where $m = -J$ for the P branch and $(J+1)$ for the R branch. The coefficient of the linear term is roughly equal to $2B$ and the coefficient of the quadratic term is equal to the difference of the B values in the two vibrational levels involved. For an allowed transition with $\Delta v = 1$, the coefficient $(B' - B'')$ equals $-\alpha_e$ and so is negative. In addition, because α_e is only a few percent of B, the magnitude of $(B' - B'')$ is very much less than B' or B''. In consequence, the vibration–rotation band consists of a series of nearly equally spaced lines with a separation of about $(B' + B'')$. This can be compared with the rotational spectrum discussed in the last chapter where the spacing is essentially constant at $2B$. The effect of the term quadratic in m in eqn 5.78 is to make the P branch diverge slightly as J increases and to make the lines in the R branch get closer together. Figure 5.14 shows the resultant spectrum. Note that the absence of a Q branch leaves a gap of approximately $4B$ between the first lines in the P and R branches (P(1) and R(0), respectively). This spacing is twice as large as that between adjacent lines in the branches and so allows the band origin to be located unambiguously.

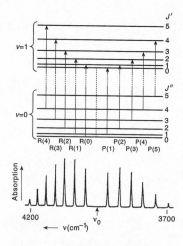

Fig. 5.14 The vibration–rotation spectrum of a diatomic molecule (actually that of HF). Note the intensity distribution in the branches and the larger gap at the band origin.

Once the quantum numbers have been attached to each line in the spectrum (a process known as *assignment*), it is desirable to determine the values of the molecular parameters ν_0, B' and B'' which model the spectrum according to eqn 5.78. This can be done either by fitting the measured wavenumbers directly to this expression or preferably by use of a method known as *combination differences*. The basis of the method is to take differences of appropriate lines in the R and P branches which share a common level in the upper or lower state. For example, what is known as the lower state combination difference is given by

$$\Delta_2 F''(J) \equiv R(J-1) - P(J+1). \tag{5.79}$$

Here Δ indicates a difference between two line positions and the subscript 2 signifies $\Delta J = 2$ for the differences. Reference to Fig. 5.15 shows that this difference depends *only* on the lower state rotational energy levels. Using the energy level expression in eqn 5.61, we can therefore calculate

$$\Delta_2 F''(J) \equiv F''(J+1) - F''(J-1)$$

$$= 2B''(2J+1) - 4D''(2J+1)(J^2 + J + 1). \tag{5.80}$$

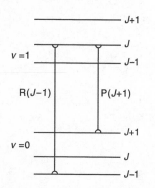

Fig. 15.15 Energy level diagram which shows how the lower state combination difference $\Delta_2 F''(J)$ is constructed in the vibration–rotation spectrum of a diatomic molecule. It can be seen that the difference between the wavenumbers of the R(J − 1) and the P(J + 1) transitions corresponds to a difference between lower state energy levels and therefore depends *only* on lower state rotational parameters.

Forming these combination differences as a function of J, dividing by $(2J + 1)$, and plotting versus $(J^2 + J + 1)$ therefore allows the B and D values for the lower level to be determined, independently of the upper-level values. In exactly the same way, we can determine the corresponding values for the upper level by forming the upper-state combination differences

$$\Delta_2 F'(J) \equiv R(J) - P(J)$$
$$= 2B'(2J + 1) - 4D'(2J + 1)(J^2 + J + 1). \qquad (5.81)$$

This separation of the dependence of the transition wavenumbers on the upper or lower state parameters is particularly useful if one or the other of the levels is *perturbed*. We shall return to this point in the next chapter but one when we discuss the structure of a vibrational band in an electronic spectrum; it is almost unheard of for the vibration–rotation spectrum of a diatomic molecule to be perturbed.

In conclusion, we have seen how a typical vibration–rotation spectrum of a diatomic molecule can be analysed (i.e. how the rotational quantum numbers can be assigned) and how the molecular constants can be determined. The constants in turn yield the force constant (k), the equilibrium bond length (r_e) and some information on the anharmonic force field (from α_e).

6 Raman spectroscopy

6.1 Introduction

We have seen how it is possible to gain structural information on diatomic molecules from their rotational and rotational–vibrational spectra (Chapters 4 and 5, respectively). In each case, energy is exchanged between the electromagnetic radiation and the molecule through the interaction between the oscillating dipole moment μ and the oscillating electric field in the radiation; the spectra are observed through *electric dipole transitions*. In consequence, there is a requirement, at least for diatomic molecules, that the molecule possess a permanent electric dipole moment, i.e. it must be heteronuclear. Information on the rotational and vibrational levels of a homonuclear diatomic molecule cannot be obtained in this way. Fortunately, there is an alternative way of obtaining the information which does not suffer from this restriction. It is called Raman spectroscopy.

Raman spectroscopy is not based on an absorption process but rather on a light scattering effect, named after the Indian physicist, C. V. Raman, who was the first to observe it in 1928 following its theoretical prediction by A. Smekal. It is easiest to appreciate the scattering process by thinking of the monochromatic light wave of frequency ν as a stream of particles, called photons, each of which carries a packet of energy $h\nu$. If this beam of photons is directed at a molecule, it will in general bounce off it (provided the frequency ν does not correspond to an absorption frequency of the molecule). The great majority of photons will be scattered in all directions but without change of energy. This process is called *elastic scattering* and was well understood by nineteenth century physicists like Lord Rayleigh; indeed it is referred to as Rayleigh scattering. A very small fraction of the photons manage to exchange some energy with the molecule as they interact with it; this is *inelastic* or *Raman scattering*. The energy is usually given to the molecule, in which case the photon moves off with energy less than $h\nu$. Under certain circumstances however, energy can also be taken from the molecule in which case the photon energy is greater than $h\nu$. We know that the internal energy of the molecule (rotation and/or vibration) is quantized. Consequently, the scattered light contains frequency components which are shifted from the incident frequency by discrete amounts.

For example, Fig. 6.1 shows a photon of energy $h\nu_{ex}$ interacting with a molecule in its zero-point level, $v = 0$. The photon can in principle raise the molecule to an energy $h\nu_{ex}$ above this level. However, if there is not a stationary state at this energy for the molecule to fetch up on, it relaxes back releasing a photon in the process. This photon may have the same energy as the incoming photon in which case we have Rayleigh scattering. More interestingly, it can have a somewhat different energy $h\nu$ allowing the molecule to fall back to the level $v = 1$ as shown in the figure. This is Raman scattering and the difference between the incoming and scattered photon energies, $h(\nu_{ex} - \nu)$, is equal to the vibrational quantum, the energy taken up by the molecule.

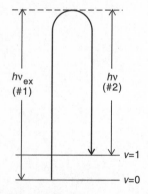

Fig. 6.1 The Raman scattering process. The incoming photon, 1, is scattered inelastically from the molecule to leave it in a higher vibrational level. Photon 2 leaves with correspondingly less energy.

In a modern Raman spectrometer, light from a monochromatic source (nowadays a laser operating at a convenient visible wavelenth) is directed at the molecular sample. The scattered light is then viewed at right angles to the incoming beam and its component wavelengths resolved with a high resolution instrument, either a monochromator or an interferometer. The arrangement is shown in Fig. 6.2. For a polyatomic molecular species, the polarization of the scattered radiation is also indicative of the type of vibration involved in a Raman peak. Since all the vibrational levels for a diatomic molecule are totally symmetric, this is not a useful diagnostic in this case.

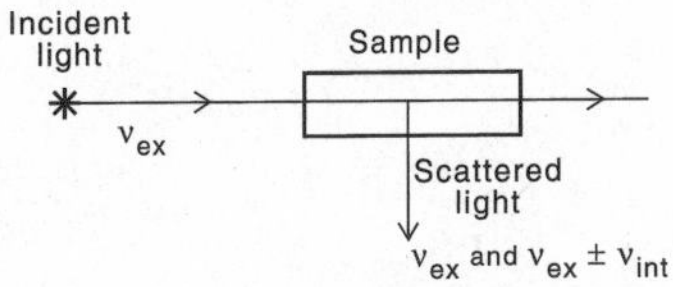

Fig. 6.2 The basic experimental arrangement for studying the Raman effect.

6.2 The nature of the interaction in Raman scattering: selection rules

In this section, we consider how energy is exchanged between the molecule and the radiation in the Raman effect. We have seen in the introduction that the scattering process is a two photon process; we call the incoming photon number 1 and the scattered photon number 2 (see Fig. 6.1). Photon 1 raises the molecule to an energy $h\nu_{ex}$ above its starting point and photon 2 is then emitted carrying away energy $h\nu$ which may be different from $h\nu_{ex}$. The intermediate energy point is sometimes referred to as a *virtual level* because the energy transferred is well-defined. However, it is important to appreciate that this does not correspond to a stationary state of the molecule (as the term *level* tends to suggest). Indeed, if it did correspond to an excited electronic state of the molecule, absorption of photon 1 would occur and completely overwhelm the scattering process.

What rôle do these two photons play at the molecular level? The first photon exposes the molecule to an oscillating electric field which causes the charges in it to move back and forth (the electrons move much more than the nuclei because of their smaller mass). As a result, there is a time-dependent electric dipole moment induced in the molecule. For low intensity light, the magnitude of of this induced dipole moment μ_{ind} is proportional to the electric field; the constant of proportionality is known as the *polarizability* α of the molecule,

$$\mu_{ind} = \alpha \, E. \tag{6.1}$$

In accordance with classical physics, this oscillating dipole moment in turn radiates electromagnetic radiation (it causes photon 2 to be emitted). The second photon does not need to take away all the initial excitation energy. The second electric dipole transition can leave the molecule in a different rotational or vibrational level, in accordance with certain selection rules which we discuss below. The important point to appreciate is that the exchange of energy between the molecule and the radiation in the Raman effect involves the polarizability and the square of the oscillating electric field:

$$W = -\tfrac{1}{2} \, \alpha \, E^2. \tag{6.2}$$

This equation is to be compared with eqn 3.2 which describes the exchange of energy for the electric dipole transitions observed in rotational or rotation–vibration spectroscopy.

It can be seen from eqn 6.2 that the interaction energy in the Raman effect is proportional to the square of the oscillating electric field. The intensity of a

Raman transition depends on the square of this interaction energy and so is proportional to E^4. The corresponding dependence for electric dipole transitions is E^2, a difference which is of course consistent with the two-photon and one-photon nature of the transitions. In practice, it means that a Raman transition is intrinsically weaker than, say, a straight rotation–vibration transition. This disadvantage of Raman spectroscopy can be overcome by raising the intensity of the exciting radiation. Fortunately, this is usually possible with present-day laser sources.

Detailed discussion of the selection rules which govern Raman spectroscopy then follows the same lines as those given in Section 3.3 for electric dipole transitions. We shall not repeat the details here but merely confine ourselves to the results.

Rotational selection rules for Raman transitions

For a molecule to show a rotational Raman spectrum (where the scattering excites or deactivates rotational motion), the polarizability α must depend on its orientation. Thus, any rotating diatomic molecule is Raman active because it has a different polarizability parallel and perpendicular to the bond; this is simply a result of its cylindrical symmetry. On the other hand, any spherically symmetric molecule, such as CH_4, is not Raman active because its polarizability is independent of orientation. The oscillating electric field in the light wave cannot couple differentially with the molecule and only Rayleigh scattering is observed.

The rotational selection rules can be derived very simply from the fact that the Raman transition is a two-photon process and involves two successive electric dipole transitions, each of which obeys

$$\Delta J = \pm 1 \text{ and } + \leftrightarrow - \tag{6.3}$$

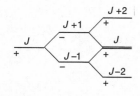

Fig. 6.3 Determination of the selection rules for Raman transitions, from their two-photon character.

for a molecule in a non-degenerate, closed-shell state (see eqns 3.22 and 3.36). Photon 1 deposits the molecule in an intermediate state with $J' = J \pm 1$ and opposite parity. Photon 2 picks up from there and takes the molecule to the final state with $J'' = J \pm 2$, J and the same parity as the original state, see Fig. 6.3. The overall rotational selection rule for Raman spectroscopy is therefore

$$\Delta J = 0, \pm 2 \tag{6.4a}$$
$$\text{parity} + \leftrightarrow + \text{ and } - \leftrightarrow -. \tag{6.4b}$$

In the unusual situation that the molecule is in a degenerate electronic state, such as nitric oxide, $\Delta J = \pm 1$ transitions are also observed. The parity selection rule however remains the same.

Vibrational selection rules for Raman transitions

The discussion of vibrational selection rules in Raman spectroscopy follows exactly the same course as that for infrared (i.e. vibrational) spectroscopy with one small change. The electric dipole operator μ must be replaced by the polarizability α since it is the property which is responsible for the exchange of energy between the molecule and the radiation field.

A general discussion of vibrational selection rules, given in Section 3.4, is applied specifically to transitions between levels of a simple harmonic oscillator in Section 5.5. The transition moment for a Raman transition between vibrational levels, the analogue of eqn 5.30, is

$$\text{transition moment} = \int \int \Psi'_{el} \Psi'_{vib} \alpha \Psi''_{vib} \Psi''_{el} d\tau_{vib} d\tau_{el}. \qquad (6.5)$$

It is now the polarizabilty α which is expanded as a Taylor series in the vibrational coordinate q:

$$\alpha = \alpha_e + (\partial\alpha/\partial q)_e q + \tfrac{1}{2}(\partial^2\alpha/\partial q^2)_e q^2 + ... \qquad (6.6)$$

In exactly the same way as for vibrational transitions, the leading term of this expansion is independent of q and so has only diagonal elements in v. Therefore it is not involved in a change of vibrational level (although it is involved in the rotational Raman effect discussed above). As before, it is the second term in the expansion, linear in q, which makes the dominant contribution. Therefore, the strongest vibrational transitions obey the same selection rules, namely $\Delta v = \pm 1$. Once again, even for a harmonic oscillator, weaker transitions conforming to $\Delta v = \pm 2$ can also be seen, arising from the third term in the expansion, quadratic in q. Thus, as in vibrational spectroscopy, there will be a strong fundamental band acompanied by weaker overtones.

These arguments tell us that, for a vibrational mode to be Raman active, the expectation value $\int \Psi_{el}(\partial\alpha/\partial q)_e \Psi''_{el} d\tau_{el}$ must be non-zero, i.e. there must be a change in the polarizability as the vibration takes place. This is a different requirement from that in vibrational (infrared) spectroscopy, even though it leads to the same selection rules for a diatomic molecule. In particular, it does not require the molecule to be polar and therefore vibrational transitions occur in Raman spectroscopy for all diatomic molecules, homonuclear as well as heteronuclear. It is reasonably easy to see whether an electric dipole moment changes during the course of a vibration. For a diatomic molecule AB, for example, the separation between the notional charges on A and B alters as the atoms move in and out. Electric polarizability is a less familiar property of a molecule and so the requirement $< \partial\alpha/\partial q > \neq 0$ is harder to envisage. Polarizability tells us about the *ease of distortion* of a collection of electric charges. In the case of a molecule, we can confine our attention to the electrons since they are much more easily displaced than the nuclei under the influence of an electric field. Some idea of how the polarizability varies as the molecule vibrates can therefore be gained by seeing how the electron cloud distorts. This is shown in Fig. 6.4. Although the discussion of vibrational selection rules is presented here for diatomic molecules, it can easily be generalized to polyatomic molecules. It is the symmetry properties of the quantity $(\partial\alpha_{ij}/\partial Q_k)$ which determine whether a particular vibration is Raman active or not. Such matters are, however, beyond the scope of this book.

6.3 Rotational Raman spectroscopy

We have seen above that it is possible to change the rotational state of a molecule through the Raman scattering process (by 2 units in J). In particular, this gives us access to the rotational levels of homonuclear molecules. In this section, we describe the appearance of a rotational Raman spectrum.

Two types of rotational transition are allowed by the selection rule $\Delta J = \pm 2$. For the first, $J+2 \leftarrow J$, energy is given to the molecule at the expense of the scattered photon whereas for the second, $J \leftarrow J+2$, the reverse happens. Since both these levels have similar populations at any reasonable

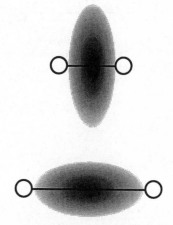

Fig. 6.4 The distortion of the electron cloud in a diatomic molecule as it vibrates, showing the variation in its polarizability.

temperature, the two types of transition will have similar intensities. The former occurs to *lower* energy of the exciting radiation (this is called a Stokes transition) while the latter occurs to *higher* energy (and is described as an anti-Stokes transition). The *shift* from the exciting energy (or wavenumber) depends only on the rotational energy levels and can be easily calculated from the formula derived in Chapter 4:

$$E_{\text{rot}} = BJ(J+1) - DJ^2(J+1)^2 + \dots \tag{6.7}$$

For the Stokes line,

$$\Delta E = F(J) - F(J+2) = -2B(2J+3) + 4D(2J+3)(J^2+3J+3). \tag{6.8}$$

For the anti-Stokes line, the shift for the transition $J \leftarrow J + 2$ is equal in magnitude but *positive*. The spacing between adjacent lines is therefore essentially $4B$, twice as large as the corresponding spacing between lines in the electric dipole rotational spectrum described in Chapter 4. The first line in this spectrum, $J = 2 \leftarrow 0$, is shifted by $6B$ from the exciting line, again in contrast to the shift of $2B$ for the $J = 1 \leftarrow 0$ line from the origin of the dipole rotational spectrum.

The intensities of the lines in the rotational Raman spectrum are described in a similar way to those in the conventional rotational spectrum (Chapter 4) although the resultant expressions are somewhat different, reflecting the different nature of the transitions. The relative intensity of a line $J + 2 \leftarrow J$ is proportional to the product of a population factor and the square of a transition moment, $N_J < \Psi_{J+2}|\alpha|\Psi_J >^2$. We note that in this case, unlike absorption spectroscopy, the intensity is proportional to the population of the initial level only, not the population difference because the transition $J \leftarrow J+2$ corresponds to a different scattering energy (anti-Stokes lines). The population factor of a given J, M_J state for a sample at a temperature T is given simply by the Boltzmann factor $\exp[-BhcJ(J+1)/kT]$. The square of the transition moment, summed over all possible connecting M_J levels, depends on J:

$$S_{J+2,J} = S_{J,J+2} = \tfrac{3}{2}(J+1)(J+2)/(2J+3). \tag{6.9}$$

This expression is to be compared with that in eqn 4.23 for ordinary dipole transitions. The variation of the line intensity with J is given by the product of $S_{J+2,J}$ and the Boltzmann factor; it is plotted out in Fig. 6.5. It can be seen that there will be a very small difference between the intensities of the corresponding Stokes ($J + 2 \leftarrow J$) and anti-Stokes transitions ($J \leftarrow J + 2$) because the population of the higher rotational level is slightly smaller. To a very good approximation therefore, the spectrum on the high energy side of the exciting wavenumber is the mirror image of that on the low energy side. In the vibrational Raman spectrum, to be discussed in the next section, the $\Delta J = +2$ and -2 transitions occur at distinct frequencies and are referred to as S and O branches respectively. The Stokes and anti-Stokes lines in the rotational spectrum are sometimes designated S and O also. However, this designation is not helpful for the rotational spectrum, because the two transitions involve the same energy levels, any more than it is helpful to think of the $\Delta J = +1$ dipole, rotational transitions as R lines.

There is one other aspect of the spectrum to consider before we can show its appearance. Because a rotational Raman spectrum can be recorded for homonuclear (as well as heteronuclear) molecules, *nuclear statistical weights*

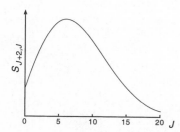

Fig. 6.5   The variation of the intensity of the $\Delta J = 2$ lines in the rotational Raman spectrum with J.

have to be taken into account. We have seen in Section 2.4 how the imposition of the Exclusion Principle on quantum systems with indistinguishable nuclei leads to the result that a significant fraction of the hyperfine levels do not exist (for a homonuclear diatomic molecule, half the levels are missing). The overall effect on individual rotational levels is that even and odd J levels have different weights because the individual nuclear hyperfine levels are not resolved. Those with the higher weight, *ortho* levels, have a weight of $(I + 1)(2I + 1)$ while those with lower weight, *para* levels, have a weight of $I(2I + 1)$. These weighting factors give the number of individual nuclear spin states for each rotational level. The ratio of the weights is $(I + 1) : I$ for *ortho* : *para* and the populations of alternate J levels will reflect this factor. If the nuclei are fermions, the level with even J will be *para* and those with odd J will be *ortho* (e.g. H_2). If the nuclei are bosons, the association is reversed (e.g. D_2). As a result, the rotational Raman spectrum of a homonuclear diatomic molecule will show an *intensity alternation* with J. For example, in the case of $^{14}N_2$, the nuclei are bosons with a spin of 1. The rotational Raman spectrum in this case shows a 2 : 1 intensity alternation. A diagram of this spectrum is shown in Fig. 6.6. For $^{16}O_2$, $I = 0$ and so alternate rotational levels (those with even N values) are missing altogether. In this situation, rotational lines are separated by $8B$. How does one distinguish between such a spectrum and that for a heteronuclear molecule with a B-value twice as large? The first line in each branch of the O_2 spectrum is $10B$ from ν_0, not $6 \times 2B$ as it would be in the other possibility. Also, the variation of the line intensity with rotational quantum number is consistent with a rotational constant B, not one twice as large.

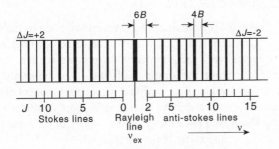

Fig. 6.6 The rotational Raman spectrum of N_2. Note the 2:1 intensity alternation.

6.4 Vibrational Raman spectrum

The strongest bands in the vibrational Raman spectrum of a diatomic molecule obey the selection rule $\Delta v = \pm 1$. The main feature is the Stokes transition, the $v = 1 \leftarrow 0$ band which lies to lower energy of the exciting line. Although the intrinsic intensity of the corresponding anti-Stokes band, $v = 0 \leftarrow 1$, is the same, the band is much weaker because the population of the $v = 1$ level is so small; indeed, for light molecules at room temperature, the population is negligible and the anti-Stokes feature is not seen at all. We shall therefore confine our attention to the Stokes, fundamental band.

The fundamental band is shifted from the exciting line by the vibrational interval ν_0, given in eqn 5.42 in the previous chapter. This vibrational band origin is equal to the harmonic vibrational wavenumber with anharmonic corrections. The rotational selection rules which operate in a vibrational transition are the same as those which govern pure rotational transitions, $\Delta J = 0, \pm 2$. The transitions which obey the selection rule $\Delta J = +2$ correspond to the largest shift from the exciting wavenumber (i.e. the scattered photons have the lowest energy) and form what is known as the S branch. Similarly, the lines which obey $\Delta J = 0$ and -2 form the Q and O branches respectively. The actual Raman shifts for individual rotational lines in each branch can be calculated from the formula for the rotational energy levels in $v = 0$ and $v = 1$, as in Section 5.8. Ignoring the effects of centrifugal distortion, the results are

$$\nu_S(J) = \nu_0 + 6B' + (5B' - B'')J + (B' - B'')J^2, \tag{6.10a}$$

$$\nu_Q(J) = \nu_0 + (B' - B'')J(J + 1) \tag{6.10b}$$

and $\quad \nu_O(J) = \nu_0 + 2B' - (3B' + B'')J + (B' - B'')J^2. \tag{6.10c}$

Thus the first line in the S branch, $J = 2 \leftarrow 0$, is $6B'$ above ν_0 whereas the first line in the O branch, $J = 0 \leftarrow 2$, lies $6B''$ below ν_0. Thereafter, the lines in each branch are spaced by roughly $4B$. Since B' is slightly smaller than B'', the S branch spacing gets smaller as J increases while, in the O branch, the lines get progressively further apart. The intensity of an individual line depends on the product of a population factor and a line strength factor, as we have just discussed for the rotational Raman spectrum. The rotational line strength factor for the S branch has been given in eqn 6.9. The corresponding factors for the Q and O branches are

$$S_{J,J} = a^0(2J + 1) + [J(J + 1)(2J + 1)/(2J + 3)], \tag{6.11a}$$

$$S_{J-2,J} = \tfrac{3}{2}J(J - 1)/(2J - 1). \tag{6.11b}$$

The quantity a^0 in eqn 6.11(a) produces an extra contribution to the scattering in the Q branch. It is technically known as trace or zeroth rank scattering. Combining these factors with the population factors, we see that the lines in each branch start out from the origin with increasing intensity, reach a maximum for a given J value, J_{max}, and then decrease exponentially beyond this value. Because $S_{J',J}$ for a Raman transition is approximately linear in J, J_{max} corresponds closely, but not exactly, to the J value of the most populated rotational level at the temperature of the sample.

The general appearance of the fundamental band in a Raman spectrum which results from these considerations is shown in Fig. 6.7. It is the spectrum of the nitrogen molecule, N_2, and so shows a $2 : 1$ intensity alternation in its branches.

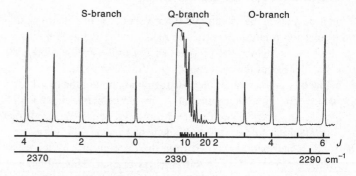

Fig. 6.7　The vibration–rotation Raman spectrum of N_2.

7 Electronic spectroscopy

7.1 Introduction

In the development of this book, the next largest quantum measured by spectroscopy involves the excitation of an electron (or strictly electrons) from one state to another. For the outermost electrons of closed-shell molecules, the amount of energy involved lies in the range between 25 000 and 50 000 cm^{-1} (or 3.1 to 6.2 eV). The corresponding wavelength region covered (400 to 200 nm) starts on the edge of the visible (violet) and goes down to the edge of the vacuum ultraviolet region; for this reason, it is sometimes called ultraviolet spectroscopy. The separation of the electronic energy levels of open-shell molecules ('free radicals') is typically quite a lot smaller. Electronic transitions in these species tend to occur at longer wavelengths and often are seen in the visible.

Each electronic state has its characteristic potential energy curve which supports its own set of vibrational and rotational levels. Electronic transitions therefore involve the change of associated vibrational and rotational quantum numbers according to selection rules which we shall describe below. First we need to define the quantum numbers of the electronic wavefunction.

7.2 Description of electronic states

The electron is a fundamental particle with a very small mass (9.109×10^{-31} kg). Consequently, its behaviour in a molecular environment requires a full quantum mechanical treatment; any attempt to understand its properties from the standpoint of classical mechanics is doomed to failure. As we have seen in Chapter 2, the properties of a molecule in a particular electronic state are all implicit in its wavefunction Ψ. This in turn is defined by the electronic Hamiltonian.

The electronic Hamiltonian

The Born–Oppenheimer separation, described in Section 2.3, encourages us to consider the motion of the electrons in a molecule separately from that of the nuclei. This amounts to keeping the nuclei fixed in space, that is, setting the nuclear kinetic energy to zero. In this situation, we can write the electronic energy as

$$W_{el} = T_{el} + V \tag{7.1}$$

where T_{el} is the electronic kinetic energy

$$T_{el} = \sum_i p_i^2 / 2m \tag{7.2}$$

and V is the potential energy arising from the electrostatic interactions between the various charged particles. For a diatomic molecule, the potential energy is given by

$$4\pi\varepsilon_0 V = -\sum_{i,\alpha} Z_\alpha e^2 / r_{i\alpha} + \sum_{i,j} e^2 / r_{ij} + Z_a Z_b e^2 / r \tag{7.3}$$

where the subscript α runs over the two nuclei a and b and i, j run over the N electrons in the molecule. The nuclear charges are $Z_a e$ and $Z_b e$. Thus $r_{i\alpha}$ is the separation between the electron i and nucleus α, r_{ij} is the separation between electrons i and j, and r is the separation between the two nuclei, i.e. the bond length (which is fixed at present in accord with the Born–Oppenheimer separation).

Transformation of this expression for the classical energy to the quantum mechanical operator form is quite straightforward, using the procedure given in Section 2.2. For each electron, the linear momentum $\boldsymbol{p}_i$ is represented by an operator

$$\boldsymbol{p}_i \rightarrow -i\hbar[\partial/\partial x_i\mathbf{i} + \partial/\partial y_i\mathbf{k} + \partial/\partial z_i\mathbf{k}] \tag{7.4}$$

so that

$$\boldsymbol{p}_i^2 \rightarrow -\hbar^2(\partial^2/\partial x_i^2 + \partial^2/\partial y_i^2 + \partial^2/\partial z_i^2) \equiv -\hbar^2\nabla_i^2. \tag{7.5}$$

The total kinetic energy is represented by the sum of N such terms. The operator representing the potential energy has exactly the same form as eqn 7.3 since it depends on the relative positions of the particles which make up the molecule and the operator which represents a position vector is itself. Thus we can write:

$$H_{\mathrm{el}} = -(\hbar^2/2m)\sum_i\nabla_i^2 + V. \tag{7.6}$$

We note in passing that we have implicitly defined the electronic coordinates in what we have written above. The instantaneous position of each electron is defined by three coordinates (x_i, y_i, z_i) measured in the molecule-fixed axis system. These coordinates do not need to be Cartesian but there will always be three of them. Therefore, for N electrons in the molecule,

$$q_{\mathrm{el}} \equiv \left\{x_1, y_1, z_1, x_2, y_2, z_2 \ldots\ldots, x_N, y_N, z_N\right\}, \tag{7.7}$$

that is $3N$ coordinates all together.

The electronic wavefunction

The electronic wavefunction which we seek is defined as the eigenfunction of the Schrödinger equation

$$H_{\mathrm{el}}\Psi_{\mathrm{el}} = E_{\mathrm{el}}\Psi_{\mathrm{el}} \tag{7.8}$$

where the operator H_{el} is defined in eqn 7.6. It can be seen from the form of eqn 7.5 that the wavefunction is an explicit function of the 3N electronic coordinates q_{el} for a given value of the bond length r. The calculation can be repeated for a different fixed value of r to give a different eigenvalue E_{el} and a different wavefunction Ψ_{el}. In this sense, r is a parameter which modifies E_{el} and Ψ_{el}. Strictly speaking, Ψ_{el} is not a function of r; rather it is a function of the electronic coordinates q_{el} but depends parametrically on r. If we plot out the parametric dependence on r, we obtain the potential energy curve introduced in Chapter 5.

So far, our progress has been deceptively straightforward. The real difficulties arise when we seek the eigenfunctions and eigenvalues of eqn 7.8. There is no known solution to this equation which puts Ψ_{el} into a simple, analytic form. Indeed, we have seen earlier that the only quantum mechanical problems which can be solved exactly are those which can be reduced to a one-body form. In the present case, even with the nuclei fixed in laboratory space, we are dealing with a $3N$-body problem so we know that, at very best, we can only obtain a numerical solution to this problem and most likely, it will be only approximately correct. The technicalities of such calculations are not the subject of this book. Rather, we need to know about the *symmetry properties* of the eigenfunctions Ψ_{el}. These can be expressed in terms of quantum numbers which in turn are used to establish the selection rules which govern transitions in electronic spectroscopy. Fortunately, we can deduce these symmetry properties from a considerably simplified and approximate form of the electronic wavefunction. We shall see how this comes about once we have introduced this approximate wavefunction.

The orbital approximation

Let us look at the form of the electronic Hamiltonian, given in eqns 7.3 and 7.6 and see how many electrons are involved in each term in the operator. The simplest of these is the nuclear–nuclear repulsion term which is independent of the electrons; indeed, it is a constant contribution to the energy in the Born–Oppenheimer approximation because r is constant. Next, there are two types of one-electron terms, the kinetic energy $-(\hbar^2/2m)\nabla_i^2$ and the electrostatic term for the attraction between each electron and the nuclei, $-Z_\alpha e^2/(4\pi\varepsilon_0 r_{i\alpha})$. Finally, there are the electron–electron repulsion terms, $+e^2/(4\pi\varepsilon_0 r_{ij})$, each of which obviously involves a pair of electrons. It is these last, two-electron terms which make the Schrödinger equation particularly difficult to solve. Recognizing this, we shall take the Nelsonian step of simply ignoring these awkward effects for the moment, in the certain knowledge that we shall have to take them back on board at a later stage. With this fairly drastic approximation, we can write the electronic Hamiltonian as a sum of one-electron operators h_i

$$H_{el} \approx H_{el}^0 = \sum_i h_i \qquad (7.9)$$

where

$$h_i = -(\hbar^2/2m)\nabla_i^2 - \sum_i Z_\alpha e^2/(4\pi\varepsilon_0 r_{i\alpha}). \qquad (7.10)$$

The advantage of this procedure is that it allows us to seek the eigenvalues and eigenfunctions of a much simpler Hamiltonian for each electron in turn, independent of the $(N - 1)$ other electrons. Our task is to solve the N independent equations of the form

$$h_i\phi_i = \varepsilon_{el}\phi_i. \qquad (7.11)$$

The eigenfunctions of these equations are one-electron wavefunctions and are known as *orbitals* (atomic orbitals if the quantum system is an atom and molecular orbitals for a molecule, as in the present case). Each orbital is a function of only three coordinates, x_i, y_i, and z_i.

The general strategy behind this approach derives from a widely used method of solving partial differential equations, the method of *separation of variables*. What we have done is to separate the dependence of the solution on the coordinates of electron i from all the other coordinates. With this realization, it is easy to see that the eigenvalues and eigenfunctions of the approximate electronic Schrödinger equation

$$H_{el}^0 \Psi_{el}^0 = E_{el}^0 \Psi_{el}^0 \qquad (7.12)$$

are given by

$$E_{el}^0 = \varepsilon_1 + \varepsilon_2 + \varepsilon_3 + ...\varepsilon_N \qquad (7.13)$$

and

$$\Psi_{el}^0 = \phi_1 \phi_2 \phi_3 ...\phi_N. \qquad (7.14)$$

In other words, the approximate value of the electronic energy is given by the sum of the individual orbital energies, with one contribution from each electron. We recognize that this is likely to be a rather poor estimate of the correct electronic energy because we have made the drastic approximation of ignoring the electron–electron repulsion terms. However, for all these limitations, the approximate eigenfunction, Ψ_{el}^0, is really useful to us because it allows us to deduce the symmetries of the various electronic states. This is because, in order to make eqn 7.14 exact, we must write

$$\Psi_{el} = \phi_1 \phi_2 \phi_3 ...\phi_N + \eta \qquad (7.15)$$

where η is a non-separated wavefunction which takes account of all the effects which we have neglected so far. Although the contribution from the correction wavefunction will be very significant, each term on the right hand side has the same symmetry, which is of course the symmetry of Ψ_{el}. Therefore, investigation of the symmetry of the orbital approximation in eqn 7.14 gives us the symmetry of the accurate wavefunction. The form of this approximation allows us to deduce these symmetry properties from those of the individual electrons which is a much simpler problem.

Molecular orbitals for diatomic molecules

We have seen in the previous sections how the formidable task of obtaining the electronic wavefunction for a molecule can be whittled down to the solution of the much simpler, one-electron Hamiltonian (eqn 7.11). There are various, well-established procedures for solving this latter problem, again beyond the scope of this book. The most common approach and perhaps the one most easily appreciated by physical chemists is the method of *linear combination of atomic orbitals* (LCAO). The basis of this approach is the assumption that the integrity of atoms is not greatly altered when they are joined together to form molecules. In other words, when an electron is in the neighbourhood of a particular nucleus in the molecule, it experiences the same interactions as it would in a free atom. In the language of quantum mechanics, these ideas are given expression by writing the molecular orbital ϕ as a linear combination of atomic orbitals χ_i,

$$\phi = \sum_i c_i \chi_i. \qquad (7.16)$$

For a diatomic molecule AB, some of these atomic orbitals are associated with atom A and the others with atom B. The relationship (eqn 7.16) is only exact when the functions $\{\chi_i\}$ form a complete basis set, which they rarely do in practice. Once again, symmetry can be used to advantage here with the requirement that ϕ can only be constructed from functions χ_i which have the same symmetry in the molecular environment.

The eigenvalues of eqn 7.11 are conveniently shown in a *molecular orbital diagram*. Let us consider the description of the electronic wavefunction for a homonuclear diatomic molecule A_2 which is composed of atoms in the first row of the Periodic Table. The molecular orbital diagram is shown in Fig. 7.1. The energies of the basis atoms are shown on each side of the diagram; these

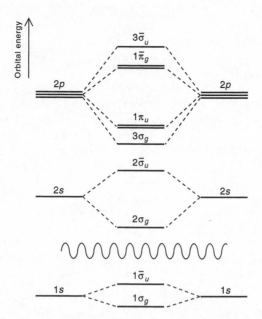

Fig. 7.1 The molecular orbital diagram for a homonuclear diatomic molecule A_2 composed of atoms from the first row of the Periodic Table. Orbitals of the same symmetry are labelled 1, 2, 3, … in order of increasing energy. The diagram is constructed assuming a fairly large separation between the $2p$ and $2s$ atomic orbitals and applies to later elements, beyond N.

levels are the same because the atoms are identical. The energy levels of the molecular orbitals which result when the two atoms interact are shown in the region in between, with connecting lines to indicate their atomic parentage. The two $2s$ orbitals combine to give a bonding σ_g and an antibonding $\bar{\sigma}_u$ orbital (the bar over the symmetry species indicates antibonding behaviour, that is, the energy of the orbital is higher than that of the separated atoms). The greater the interaction between the $2s$ orbitals (or 'overlap' as it is sometimes described), the larger the energy difference between the σ_g and $\bar{\sigma}_u$ orbitals. In the cylindrical environment of the diatomic molecule, the $2p$ orbitals form representations of σ $(2p_z)$ and π $(2p_{x,y})$ symmetry. These can be formed into sum and difference combinations which have a definite symmetry under the inversion operator, i. Like the $2s$ orbitals, the $2p_z$ orbitals give σ_g bonding and $\bar{\sigma}_u$ antibonding orbitals with, generally, a larger separation because their directional properties lead to greater overlap. The $2p_{x,y}$ orbitals form a bonding π_u orbital and an antibonding $\bar{\pi}_g$ orbital. The different g,u association follows simply from the form of these orbitals (see Fig. 7.2). The energy separation of the π orbitals is smaller because the relative orientation of its component $2p$ orbitals does not permit large overlap.

It will be noticed in the molecular orbital diagram in Fig. 7.1 that there are multiple occurrences of orbitals of a particular symmetry, such as σ_g. These orbitals are distinguished from each other by labelling them with a number which indicates their position in an ordering scheme of increasing energy. Thus the $2\sigma_g$ orbital is the second highest orbital of σ_g symmetry (there is one σ_g orbital below it in energy which arises from the in-phase combination of $1s$ orbitals on each atom). So far, we have described the construction of the molecular orbital diagram without considering the possible mixing of orbitals

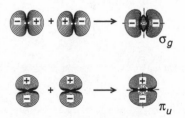

Fig. 7.2 The formation of σ_g and π_u orbitals from $2p$ atomic orbitals in a homonuclear diatomic molecule.

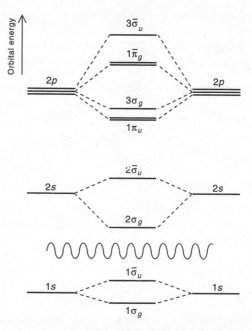

Fig. 7.3   The molecular orbital diagram for a homonuclear diatomic molecule A_2 for earlier elements of the first row of the Periodic Table, up to and including N_2. The small $2p$–$2s$ separation causes large shifts of orbitals of the same symmetry.

arising from the $2s$ or $2p$ atomic orbitals. Clearly, such mixing can occur. For example, the $2\sigma_g$ and $3\sigma_g$ orbitals have the same symmetry and will interact, the lower orbital being pushed down and the upper raised in energy. The closer the $2p$ and $2s$ orbitals, the greater these energy shifts. It is well-known that the $2p$–$2s$ separation gets larger across the first row in the Periodic Table. Therefore, the interaction between the σ_g orbitals (or the σ_u orbitals) gets smaller as the atomic number increases. However, for early members of the first row, the resultant shifts are much larger, so much so in fact that the $3\sigma_g$ orbital gets pushed *above* the π_u orbital (whose energy is unaffected because there are no other orbitals of π_u symmetry with $n = 2$). The $3\bar{\sigma}_u$ orbital is also raised in energy at the expense of the $2\bar{\sigma}_u$ orbital, thereby increasing its separation from the $1\bar{\pi}_g$ orbital. This modified form of the molecular orbital diagram is shown in Fig. 7.3. It is applicable to atoms from the first row up to and including N; beyond this, the ordering of orbitals is that shown in Fig. 7.1.

The molecular orbital diagram we have considered so far is that for a homonuclear diatomic molecule. What happens when we are dealing with a heteronuclear species, AB? In this case, the atom with the higher nuclear charge will hang on to its electrons more firmly, i.e. it is more electronegative. The diagrams in Figs 7.1 and 7.3 can be modified quite simply so that they describe the molecular orbitals of AB reliably. Let us assume that B is more electronegative than A. The atomic orbital energies of B are then lower than the corresponding orbital energies of A. The molecular orbitals are then constructed in the same way except for the following differences:

1. The molecular orbitals are no longer 50 : 50 mixtures of the orbitals on each atom. The lower molecular orbital of a pair will reflect more of the atomic

orbital on the more electronegative atom, B and the upper one of the pair will have a larger contribution from A. In the extreme case of a molecule consisting of two atoms with vastly different electronegativities, the lower molecular orbital will be an almost completely pure atomic orbital on B.

2. Since the molecule AB does not have a centre of symmetry, g and u are no longer appropriate symmetry labels. These are therefore dropped from the symmetry designations. Furthemore, orbitals which are of different symmetries in A_2 (such as σ_g and σ_u) now belong to the same irreducible representation and so can push each other around in the molecular orbital diagram. Thus, in Fig. 7.1 for example, all four σ orbitals are mixed together to a greater or lesser extent and it is much harder to predict the relative positions of these orbitals.

These features are all collected together in Fig. 7.4, which shows the molecular orbital diagram for a diatomic molecule composed of atoms A and B in the first half of the Periodic Table.

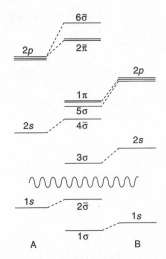

Fig. 7.4 The molecular orbital diagram for a heteronuclear diatomic molecule AB for earlier elements of the first row of the Periodic Table.

Electron configurations: the Aufbau Prinzip

We have reached the point where we have determined all the requisite orbitals and orbital energies (i.e. the solutions to eqn 7.11). How do we now estimate the electronic energy of a particular molecular species and also determine the overall symmetry of its electronic states? The procedure, called the *aufbau prinzip* (or the building-up principle), is quite simple and is as follows: The electrons from the two atoms are assigned to the available orbitals, two at a time and in order of increasing energy. This is the procedure for the construction of the ground state; excited electronic states are obtained by promotion of one or more electrons from a lower to a higher molecular orbital. The total energy of the state is estimated by adding a contribution ε_i for each electron in accord with eqn 7.13. The absolute energy obtained in this way is almost certainly very unreliable but energy differences estimated on the promotion of particular electrons may not be quite so bad.

No more than two electrons can be assigned to a particular orbital and they must have their spins paired (i.e. one is in an α or $m_s = 1/2$ spin state and the other in a β or $m_s = -1/2$ state). This is in order to satisfy the Pauli exclusion principle, which we met earlier on in Section 2.4. The principle decides which states are permissible on the basis of the interchange of any pair of identical particles. In Chapter 2, we were considering the interchange of identical nuclei in a molecule. Exactly the same principle applies to the permutation P_{12} of pairs of electrons, all of which are indistinguishable in a molecule. Because the electron spin is 1/2, electrons obey Fermi–Dirac statistics and

$$P_{12}\Psi_{el} = -\Psi_{el}. \tag{7.17}$$

There is a slight complication here because the full electronic wavefunction depends on both orbital and spin properties of the electron. The interaction between these two angular momenta in a molecule is usually quite small so we can think of these characteristics separately. It can be seen that, if we assign exactly the same orbital and spin quantum numbers to two different electrons in a molecule and then permute this pair under P_{12}, the wavefunction would be completely unaffected and so eqn 7.17 would not be satisfied. Thus, another way of stating the Pauli exclusion principle for electrons is to say that no two electrons in a molecule can have the same set of quantum numbers. Thus, after we have put one electron into a particular orbital with spin α, say, the second electron must go in with opposite spin β for the Pauli principle to be satisfied. Furthermore, we cannot put more than two electrons into an orbital because there are only two distinct spin states for an electron.

The approximate electronic wavefunction Ψ_{el}^0 is constructed from the molecular orbitals according to eqn 7.14. Strictly, it is composed of the product not just of the occupied orbitals but of the occupied *spin-orbitals*. Thus, for the simplest diatomic molecule in the first row of the Periodic Table, Li_2, the wavefunction can be written symbolically as

$$\Psi_{el} = K\, 2\sigma_g^\alpha 2\sigma_g^\beta \tag{7.18}$$

where the superscript gives the electron spin; that is, the two outermost electrons are both assigned to the lowest $2\sigma_g$ orbital with spins paired. The spin-orbital in eqn 7.18 only refers explicitly to the valence shell electrons. There are also four other electrons in Li_2 in the $1\sigma_g$ and $1\bar{\sigma}_u$ orbitals which arise from the $1s$ atomic orbitals. These electrons are much lower in energy and are not involved in the excitation process to give the first few electronic states of Li_2. Their presence is summarized by the symbol K which indicates that there is a filled K shell in this case. In a similar manner, we can write the configuration for N_2 in its electronic ground state,

$$\Psi_{el} = K\, 2\sigma_g^\alpha 2\sigma_g^\beta 2\bar{\sigma}_u^\alpha 2\bar{\sigma}_u^\beta 1\pi_{u,+1}^\alpha 1\pi_{u,+1}^\beta 1\pi_{u,-1}^\alpha 1\pi_{u,-1}^\beta 3\sigma_g^\alpha 3\sigma_g^\beta. \tag{7.19}$$

Note here that we distinguish between the two degenerate π molecular orbitals by giving the value of l_z (1 and -1); this is the component of the electronic orbital angular momentum along the molecular axis. It can be seen that once a molecular orbital contains its two electrons (the so-called closed shell), we no longer need to specify the electron spins. Thus the orbital wavefunction for N_2 (given in eqn 7.20), is usually simplified to

$$\Psi_{el} = K(2\sigma_g)^2(2\bar{\sigma}_u)^2(1\pi_u)^4(3\sigma_g)^2. \tag{7.20}$$

This short-hand description of the electronic wavefunction is called an *electronic configuration*. If all the orbitals contain their full complement of electrons, the configuration is referred to as a *closed-shell configuration*. For a very good reason, almost all chemically stable molecules have a closed-shell ground configuration.

Open-shell configurations are of course possible and, in a few cases, the molecules are stable in their ground electronic states. Well-known examples are molecular oxygen and nitric oxide. For example, the ground configuration of O_2 is

$$\Psi_{el} = K(2\sigma_g)^2(2\bar{\sigma}_u)^2(3\sigma_g)^2(1\pi_u)^4(1\bar{\pi}_g)^2. \tag{7.21}$$

The outermost, anti-bonding π orbital is only half full in this case. It is possible for several electronic states to arise in such a situation as we shall see in the following section.

Electronic states; quantum numbers for the electronic wavefunction

The electronic configurations can be used to derive the nature of the electronic states Ψ_{el} which exist for that molecule. There are two aspects, the total electron orbital momentum and electron spin momentum. The electronic orbital angular momentum for each electron $\mathbf{l}_i$ can be added vectorially to give the total angular momentum $\mathbf{L}$

$$\mathbf{L} = \mathbf{l}_1 + \mathbf{l}_2 + \mathbf{l}_3 + ...\mathbf{l}_N. \tag{7.22}$$

However, although $\mathbf{l}_i$ and $\mathbf{L}$ are well-defined quantities in the spherically symmetric surroundings of an atom, in the cylindrical environment of a diatomic molecule, they are not. This difference can be appreciated from the realization that the distribution of charges in a diatomic molecule produces an axially symmetric electric field. For an atom in such a field, a Stark effect occurs; there is a shift and possibly a splitting of the various m states involved. Consider, for example, an atom in a $l = 1$ state (a p orbital); in an electric field, the $m_l = 0$ component is shifted from the $m_l = \pm 1$ pair which remain degenerate in the field. These effects occur because the electric field mixes states with different l values. In the limit of heavy mixing, l ceases to be a good quantum number. However, the component of $\mathbf{l}$ along the field direction, m_l, remains a good quantum number. The field direction in a diatomic molecule lies along the internuclear axis. Because m_l is usually taken to define the component of the orbital angular momentum along a laboratory-fixed axis, Z, we shall use the symbol λ_i to indicate the component along the molecule-fixed axis z. Therefore, although eqn 7.22 is not meaningful for a diatomic molecule, its component along the z axis is, namely:

$$\Lambda = \lambda_1 + \lambda_2 + \lambda_3 + ...\lambda_N. \tag{7.23}$$

Here Λ is the component of the total orbital angular momentum and is one of the quantum numbers used to characterize Ψ_{el}. It can be determined simply from the algebraic sum of the individual λ values, remembering that, for a closed shell, $\Sigma_i \lambda_i$ is zero.

In addition, each electron carries a spin angular momentum of 1/2. Thus the total spin angular momentum $\mathbf{S}$ is the vector sum of all these individual vectors:

$$\mathbf{S} = \mathbf{s}_1 + \mathbf{s}_2 + \mathbf{s}_3 + ...\mathbf{s}_N. \tag{7.24}$$

Now the cylindrical symmetry of a diatomic molecule imposes itself in three-dimensional space but not in the rather mysterious space spanned by spin coordinates. Thus, even in a molecular environment, $\mathbf{S}$ remains a well-defined quantity. (It only ceases to be so in molecules containing very heavy atoms for which spin-orbit coupling effects mix states with different S values.) The electron spin quantum numbers for the electronic wavefunction of a diatomic molecule are therefore S and its component along the molecule-fixed z axis, which is called Σ where

$$\Sigma = S, S - 1, S - 2, ... - S. \tag{7.25}$$

Once again, the evaluation of the allowed values for S for any particular configuration is considerably simplified on the realization that the spin angular momentum of the electrons in a closed shell is zero, because all the spins must be paired in accordance with the Pauli exclusion principle.

We are now in a position to be able to work out the electronic quantum numbers for the electronic ground state of N_2, from the configuration in eqn 7.20. This is a particularly simple exercise since all the orbitals have a closed shell structure. Therefore, both the orbital angular momentum Λ and the spin angular momentum are zero:

$$S = 0, \quad \Lambda = 0. \tag{7.26}$$

This result is quite general for any closed-shell configuration.

This symmetry description of the electronic wavefunction is summarized in what is known as the *spectroscopic term*, written as $^{2S+1}\Lambda_{\Omega}$. The main symbol gives the Λ value. Traditionally, the symbol Σ is used for $\Lambda = 0$, Π for $\Lambda = \pm 1$, Δ for $\Lambda = \pm 2$, etc.. The left superscript is called the *multiplicity* of the state and gives the number of spin components for that state. Each of these spin components is labelled by the appropriate value for the quantum number Ω which appears as a lower right subscript to the main symbol. Ω is the component of the total electronic angular momentum, spin and orbital, along the internuclear axis. Thus

$$\Omega = \Lambda + \Sigma. \tag{7.27}$$

The spectroscopic term for the ground state of N_2 is therefore written $^1\Sigma_g^+$. The multiplicity is 1, that is, there is no spin degeneracy. The main symbol is Σ_g^+ which is the totally symmetric representation in the point group for N_2, $D_{\infty h}$. Quite generally, the orbital characteristic of a closed-shell state of any molecule is the totally symmetric representation of the appropriate point group. The value for Ω, i.e. zero, is usually omitted from the spectroscopic term for a $^1\Sigma_g^+$ closed shell state because its value is obvious.

What happens if the configuration is open shell? Let us consider the case of molecular oxygen, O_2, whose ground configuration is given in eqn 7.21, $K(2\sigma_g)^2(2\bar{\sigma}_u)^2(3\sigma_g)^2(1\pi_u)^4(1\bar{\pi}_g)^2$. We can ignore the closed-shell orbitals because they only make a totally symmetric contribution to the orbital wavefunction and no contribution to the total spin angular momentum. We therefore only have to consider the two electrons in the $\bar{\pi}_g$ orbital. Let us consider the orbital character first. Each electron can occupy either the π orbital with $\lambda = +1$ or the orbital with $\lambda = -1$; there are four possible arrangements:

$$\phi_{+1}(1)\phi_{+1}(2) \qquad \text{i.e. } \Lambda = +2 \tag{7.28a}$$
$$\phi_{+1}(1)\phi_{-1}(2) \qquad \text{i.e. } \Lambda = 0 \tag{7.28b}$$
$$\phi_{-1}(1)\phi_{+1}(2) \qquad \text{i.e. } \Lambda = 0 \tag{7.28c}$$
$$\phi_{-1}(1)\psi_{-1}(2) \qquad \text{i.e } \Lambda = -2. \tag{7.28d}$$

It is easy to see that the permutation operator P_{12} leaves the products in eqns 7.28(a) and 7.28(d) unaltered; in other words, they are symmetric with respect to P_{12}. These wavefunctions form the two degenerate components of a Δ_g state with $\Lambda = \pm 2$. The other two combinations (eqns 7.28(b) and 7.28(c)), are indeterminate because P_{12} transforms them into different functions:

$$P_{12}\phi_{+1}(1)\,\phi_{-1}(2) = \phi_{-1}(1)\,\phi_{+1}(2) \tag{7.29a}$$
$$P_{12}\phi_{-1}(1)\,\phi_{+1}(2) = \phi_{+1}(1)\,\phi_{-1}(2). \tag{7.29b}$$

These product functions are therefore not eigenfunctions of P_{12} (or, in the language of group theory, they are not irreducible representations of the permutation group S_2). However, since they transform into each other under P_{12}, we have only to take sum and difference combinations for them to have well-defined behaviour under P_{12}:

$$P_{12}[\phi_{+1}(1)\phi_{-1}(2) \pm \phi_{-1}(1)\phi_{+1}(2)] = \\ \pm [\phi_{+1}(1)\phi_{-1}(2) \pm \phi_{-1}(1)\phi_{+1}(2)]. \tag{7.30}$$

In other words, the + combination is symmetric with respect to P_{12} and the − combination is antisymmetric. Both these combinations form Σ states; it turns out that the + combination corresponds to a Σ_g^+ state and the − combination to Σ_g^- (both are *gerade* states because each of the orbitals is g and $g \times g = g$). From the orbital point of view, therefore, we have Δ_g and Σ_g^+ functions, both of which are symmetric and a Σ_g^- function which is anti-symmetric with respect to P_{12}.

The classification of the electron spin functions with respect to permutation of the two electrons is dealt with in a similar fashion. Indeed, we have already solved this problem for two identical spin 1/2 particles in Section 2.4, although in that case we were dealing with two protons rather than two electrons. The conclusion is nevertheless the same. Three spin functions are symmetric with respect to P_{12}

$$P_{12}\alpha\alpha = \alpha\alpha \tag{7.31a}$$

$$P_{12}1/\sqrt{2}(\alpha\beta + \beta\alpha) = 1\sqrt{2}(\alpha\beta + \beta\alpha) \tag{7.31b}$$

$$P_{12}\beta\beta = \beta\beta \tag{7.31c}$$

and form a triplet set; the other is antisymmetric and is a singlet:

$$P_{12}1/\sqrt{2}(\alpha\beta - \beta\alpha) = -1/\sqrt{2}(\alpha\beta - \beta\alpha). \tag{7.32}$$

In order to form the electronic wavefunction Ψ_{el} which is antisymmetric with respect to P_{12}, we must combine either the symmetric spin function with the antisymmetric orbital function ($^3\Sigma_g^-$) or the antisymmetric spin function with the symmetric orbital function ($^1\Delta_g$ and $^1\Sigma_g^+$). Thus, only three electronic states arise from the ground configuration of O_2; the other three possibilities ($^1\Sigma_g^-$, $^3\Delta_g$ and $^3\Sigma_g^+$) are not allowed by the Pauli exclusion principle. The three allowed states lie close in energy because they arise from the same configuration. Their relative ordering can be deduced by applying Hund's rules (even though these rules are intended for atomic systems, not molecular). The lowest state is $^3\Sigma_g^-$ because it has the highest multiplicity; this then is the ground state of O_2. Of the other two states, both of which are singlets, the $^1\Delta_g$ is lower because it has the higher orbital angular momentum along the axis. The relative positions of these states are summarized in Fig. 7.5.

Another example of a molecule with an open shell electronic structure is the CN radical; unlike O_2, this is not chemically stable but is a reactive, short-lived molecule. The appropriate molecular orbital diagram is given in Fig. 7.4. Filling up in accord with the Aufbau Prinzip, we obtain the ground configuration

$$\Psi_{el} = K(3\sigma)^2(4\bar{\sigma})^2(1\pi)^4(5\sigma)^1. \tag{7.33}$$

There is only one electron in the open shell orbital so it is easy to see that $S = 1/2$, $\Lambda = 0$, i.e. the ground state is $^2\Sigma^+$. We can see in Fig. 7.4 that the 1π and 5σ orbitals lie very close together. Very little energy is therefore required to promote an electron from the π to the σ orbital to give the first excited state

$$\Psi'_{el} = K(3\sigma)^2(4\bar{\sigma})^2(1\pi)^3(5\sigma)^2. \tag{7.34}$$

Three electrons in the π orbital correspond to one 'hole' in the π orbital. Thus the state arising from this configuration is a $^2\Pi$ state. Because the open-shell orbital is more than half-full, the spin-orbit splitting in this state is inverted, that is, the $^2\Pi_{3/2}$ component lies below the $^2\Pi_{1/2}$ component.

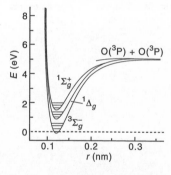

Fig. 7.5 The electronic states which arise from the ground configuration of molecular oxygen, ... $(1\pi_u)^4(1\bar{\pi}_g)^2$. These are the only states permitted by the Pauli principle.

7.3   The vibrational and rotational structure of electronic states

The motion of the electrons in a molecule takes place on a very short timescale of the order of 10^{-15} s (or 1 fs), for the duration of which the nuclei are fixed. Over a longer time period than this, we have also to consider the effects of nuclear motion. We shall first consider vibrational motion (10^{-13} s) and then rotational motion (10^{-11} s).

Vibrational energy levels

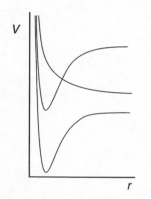

As we have described in Chapter 5, the parametric dependence of the electronic energy of a diatomic molecule on the internuclear distance r can be plotted out to give the potential energy curve $V(r)$, see Fig. 7.6. Such a potential energy curve exists for every electronic state of the molecule, each with its characteristic r_e value, force constant k (or the curvature at the minimum), and dissociation energy D_e. These characteristics can vary considerably from one state to another. Indeed, some potential energy curves may not even show a minimum and are said to be purely repulsive (see also Fig. (7.6)). A diatomic molecule has at best only a transitory existence in a repulsive state. Such states are difficult to study directly by spectroscopy but are not completely without interest as we shall see later.

Fig. 7.6   Some typical potential energy curves of a diatomic molecule.

In the same way as we saw for the ground electronic state in Chapter 5, each bound potential energy curve supports its own set of vibrational energy levels

$$G(v) = E_{\text{vib}}/hc = (v + \tfrac{1}{2})\omega_e - (v + \tfrac{1}{2})^2 \omega_e x_e + ... \qquad (7.35)$$

The vibrational parameters ω_e, $\omega_e x_e$, ... are characteristic of the particular state and can be used as a source of information on the structural properties of the molecule in that state. The stronger the bond between the atoms, the larger the dissociation energy. This usually means that the vibrational frequency (or wavenumber) will be higher also although there is not a rigorous connection between these two properties (the force constant k and the dissociation energy D_e are independent properties).

Rotational energy levels

For a time interval longer than about 100 ns, a molecule in a particular vibrational level of a given electronic state will also show the effects of quantized rotational motion as we have discussed in Chapter 4. As a result, there is an additional contribution to the energy of the form

$$F(J) = E_{\text{rot}}/hc = B_v J(J + 1) - D_v J^2(J + 1)^2 + ... \qquad (7.36)$$

The rotational constant B_v depends weakly on the vibrational quantum number v according to

$$B_v = B_e - \alpha_e(v + \tfrac{1}{2}) + ... \qquad (7.37)$$

Since the vibrational dependence parameter α_e is positive for an average over a symmetric stretching vibration, the B_v value gets smaller as v increases. (The centrifugal distortion parameter D_v also shows a weak vibrational dependence but such effects generally show up only for very precise measurements.)

We recall that the rotational constant depends inversely on the square of the bond length r:

$$B_e = \hbar^2/(2\mu r_e^2 hc). \qquad (7.38)$$

Since the equilibrium bond length r_e can change considerably from one electronic state to another (because of a difference in bonding), the B value can also be significantly different. This is in contrast to the case of vibrational excitation where, as we have noted above, the B-value changes only slightly. The changes on electronic excitation are usually large enough for the B-value to be characteristic of a particular electronic state.

7.4   Transitions in electronic spectroscopy: selection rules

Having established the energy level scheme for a rotating, vibrating molecule in an electronic state, we now investigate the selection rules which govern spectroscopic transitions between these states. We have already discussed the derivation of these rules in some detail in Chapter 3. Here we shall simply summarize the conclusions.

The selection rules for the electronic quantum numbers Λ and S reflect the symmetry properties of the electronic transition moment $\int \Psi_{el}'' \mu_\alpha^e \Psi_{el}'' d\tau_{el}$, where α is a molecule-fixed component of the electric dipole moment μ_α^e. The rules are:

$$\Delta\Lambda = 0, \pm1, \qquad (7.39a)$$

$$\Delta S = 0, \qquad (7.39b)$$

$$g \leftrightarrow u. \qquad (7.39c)$$

For the first of these selection rules, the transition $\Delta\Lambda = 0$ is induced by the z component of μ while that with $\Delta\Lambda = \pm1$ is induced by the x,y component of μ. The electron spin selction rule, eqn 7.39(b), is not a strict condition; significant departures from it can be seen when spin-orbit coupling effects become important as they do with heavier atoms. There is one other selection rule which applies to $\Delta\Lambda = 0$ transitions between Σ states. The component of the dipole moment operator along the z axis does not alter the behaviour of the wavefunction with respect to reflection in any plane which contains the axis. Thus only the transitions $\Sigma^+ - \Sigma^+$ and $\Sigma^- - \Sigma^-$ are allowed.

The vibrational selection rules for an electronic transition depend on the vibrational overlap factor $\int \Psi_{vib}' \Psi_{vib}'' d\tau_{vib}$. Because the wavefunction for any vibrational level of a diatomic molecule is totally symmetric (Σ^+), there is in fact no symmetry restriction on the change in the v quantum number. The relative intensity of any particular transition (v', v'') depends on the square of this overlap integral, a quantity which is known as the Franck–Condon factor. We shall discuss what makes this factor large or small in the next section.

The rotational selection rules are derived from the properties of the direction cosine matrix elements, $\int \Psi_{rot}' \lambda_{Z\alpha}(\theta, \phi, \chi) \Psi_{rot}'' d\tau_{rot}$. They are the usual electric dipole selection rules

$$\Delta J = 0, \pm1; \; + \leftrightarrow -. \qquad (7.40)$$

Therefore, we expect in general to see three branches of lines, P, Q and R corresponding to the $\Delta J = -1$, 0 and $+1$, respectively. If the electronic transition is between Σ states, i.e. $\Sigma^\pm - \Sigma^\pm$, the $\Delta J = 0$ transitions are forbidden by the parity selection rule in exactly the same way as they are for vibration–rotation transitions of closed-shell diatomic molecules (see Section 5.8). For a diatomic molecule, the line strength factors, $S_{J'J''}$, are given by the

Table 7.1　Hönl – London line-strength factors[a]

$\Delta\Lambda = 0$	$S_{J+1,J} = (J+1+\Lambda)(J+1-\Lambda)/(J+1)$
	$S_{J,J} = (2J+1)\Lambda^2/[J(J+1)]$
	$S_{J-1,J} = (J+\Lambda)(J-\Lambda)/J$
$\Delta\Lambda = \pm 1$	$S_{J+1,J} = (J+2\pm\Lambda)(J+1\pm\Lambda)/[4(J+1)]$
	$S_{J,J} = (J+1\pm\Lambda)(J\mp\Lambda)(2J+1)/[4J(J+1)]$
	$S_{J-1,J} = (J-1\mp\Lambda)(J\pm\Lambda)/[4J]$

[a] J and Λ are the quantum numbers for the lower state

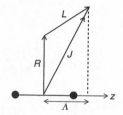

Fig. 7.7　The coupling of the rotational (**R**) and electronic (**L**) angular momenta in a diatomic molecule. Because **R** is perpendicular to the internuclear axis, both **J** and **L** have the same projection on the z axis.

Hönl–London expressions which are reproduced in Table 7.1. Note that the intensities of the rotational transitions differ, depending on whether $\Delta\Lambda = 0$ or ± 1. For $\Delta\Lambda = 0$, the Q branch is strongest at low J and falls off rapidly as J increases; for $\Delta\Lambda = \pm 1$, the Q branch line–strength factor increases linearly with J. The factor $S_{J'J''}$ is $\propto \Lambda^2$ for the Q branch when $\Delta\Lambda = 0$ and therefore vanishes for a $\Sigma - \Sigma$ transition, as stated above. An important result follows from this dependence of the rotational intensity on the $\Delta\Lambda$ selection rule. Analysis of the rotational structure of the electronic transition allows the $\Delta\Lambda$ selection rule to be determined for any particular transition (indeed, it allows the symmetries of the upper and lower electronic states to be determined); hence the importance of recording a spectrum at rotational resolution. The reason for the dependence of the rotational intensities on the electronic quantum number Λ is that the total angular momentum J is the vector sum of the nuclear, end-over-end rotation R and the orbital angular momentum L, as shown in Fig. 7.7. Because R is perpendicular to the internuclear axis,

$$J_z = L_z. \tag{7.41}$$

Hence, the component of the orbital angular momentum Λ along the z axis is also the z component of the total angular momentum J. Thus Λ is involved in the description of the rotational as well as the electronic wavefunctions.

7.5　The vibrational (band) structure of an electronic spectrum

The vibrational structure: the Franck–Condon principle

We have seen in Chapter 3 that there is no restriction on the change in v for a vibrational transition but that the intensity of a particular band depends on the Franck–Condon factor. Let us consider an electronic spectrum recorded in absorption, in other words all the vibrational transitions start from the zero-point level $v'' = 0$. The wavefunction for this level has been given in Fig. 5.7. The strongest vibrational bands will involve those upper state levels with the largest overlap with this wavefunction. In consequence, the upper state wavefunctions must have a large magnitude at $r = r_e$. If the potential energy curve for the upper state is very similar to that for the lower state, including the equilibrium bond length, much the largest overlap will occur for the upper state vibrational level $v' = 0$. In this situation, the (0, 0) band will completely dominate the spectrum. If, on the other hand, the potential energy curve for the upper state is different in its properties and, in particular, its equilibrium bond

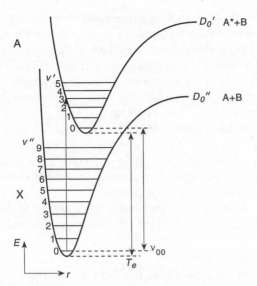

Fig. 7.8 The vibrational transition in an electronic transition of a diatomic molecule which is most favoured by the Franck–Condon principle.

length is larger (or smaller) than that of the ground state, the vibrational level with large amplitude at $r = r_e''$ will be considerably excited. For all but the lowest few vibrational levels, the amplitude of the vibrational wavefunction is largest near the classical turning points for vibrational motion. Therefore, the most probable vibrational transition can be predicted by drawing a vertical line at $r = r_e''$ on the energy level digram to see where it intersects the upper state potential energy curve (see Fig. 7.8). In this situation, the transitions to the upper state vibrational levels on either side of the most favoured one will also have significant intensity. In consequence, we can appreciate a quite general statement on the intensity distribution of vibrational transitions in electronic spectroscopy. If there is a change of equilibrium bond length on electronic excitation, the intensity of the transition will be divided among a run of adjacent vibrational bands . . $(v+2, 0), (v+1, 0), (v, 0), (v-1, 0), (v-2, 0)$, . . which is known as a *progression*. The general appearance of a vibrational progression is sketched in Fig. 7.9; it consists of a series of roughly equally spaced bands at intervals of approximately ω_e'.

The Franck–Condon principle has been discussed above from a quantum mechanical point of view. There is also a classical justification for the principle which states that, because the electrons move very much faster than the nuclei in a molecule, a transition between the electronic states occurs in such a short time interval that the nuclei do not alter their positions or momenta significantly. Thus the electronic transition can be represented by a vertical line (at constant r) on the potential energy diagram—as we have already suggested in Fig. 7.8. This description of the Franck–Condon principle leads to a widespread misunderstanding of the processes involved in an electronic transition. The argument goes as follows: the Einstein A coefficient for an electronic transition around 400 nm has a magnitude of about 10^8 s^{-1}, that is the transition between electronic states occurs in about 10^{-8} s. However, a typical vibrational period is 10^{-13} s which is considerably *shorter* than the electronic transition time, as estimated above. This appears to

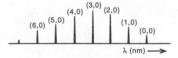

Fig. 7.9 A vibrational progression in an electronic transition of a diatomic molecule between two states with different r_e values, as shown in Fig. 7.8.

undermine the classical arguments presented to justify the Franck–Condon principle—indeed it even threatens the Born–Oppenheimer separation!

The source of confusion here is in the use of the phrase 'the time taken for an electronic transition'. This description is at best misleading and, at worst, simply incorrect. The time interval that is relevant is actually the period of an electron in the orbit associated with the electronic transition. It is important to realize that the classical picture of an electron in an orbit is obtained from the time-dependent wavefunction which results from the combination of two stationary state wavefunctions between which the transition is observed. The period of the electron in its orbit is just the reciprocal of the transition frequency. The correct classical statement of the Franck–Condon principle (or the Born–Oppenheimer separation for that matter) is that the nuclear positions and momenta do not change appreciably in the period of the electron in its orbit which is typically much shorter than the interaction time, around 10^{-15} s or 1 fs.

The rotational (or branch) structure in an electronic transition

Each vibrational band in an electronic spectrum consists of a number of rotational lines. The description of this rotational structure involves very similar considerations to those already described for the branch structure in a vibration–rotation spectrum in Section 5.8.9. The wavenumbers of the individual lines in the P, Q, and R branches are given by

$$\nu_R(J) = \nu_0 + (3B' - B'')J + (B' - B'')J^2 \tag{7.42a}$$

$$\nu_Q(J) = \nu_0 + (B' - B'')J(J + 1) \tag{7.42b}$$

$$\nu_P(J) = \nu_0 - (B' + B'')J + (B' - B'')J^2 \tag{7.42c}$$

where we have ignored centrifugal distortion effects. The band origin ν_0 has contributions from the electronic and vibrational energies of the two states

$$\nu_0 = E'_{el} - E''_{el} + G'(v') - G'(v'') \tag{7.43}$$

where the single and double primes label the upper and lower states, respectively. The intensities of the lines in each branch are given by

$$I_{rel} \propto S_{J'J} \exp[-hcB''J(J+1)/kT] \tag{7.44}$$

where the line strength factors $S_{J'J}$ are given by the Hönl–London expressions in Table 7.1. As in the case of vibration–rotation or Raman spectra, the Boltzmann factor causes the intensity of the lines in each branch to fall off exponentially at high J values. In the P and R branches, the strongest lines occur for J values near the most highly populated rotational level for which

$$J_{max} = \tfrac{1}{2}[(2kT/Bhc)^{1/2} - 1]. \tag{7.45}$$

Despite the similarities in their occurrence, the rotational branch structure in an electronic spectrum can look very different from that observed in a vibration–rotation band. This is because the B values for the two states can now differ by a much larger amount; the magnitude of $(B' - B'')$ can be as much as a fifth of $\tfrac{1}{2}(B' + B'')$ rather than the 2 or 3 per cent as it is for vibration–rotation transitions. As a result of this, the quadratic coefficient of J in eqn 7.42 is not much smaller than the linear coefficient. Let us consider the case when $B' < B''$. For infrared bands, this causes the lines in the R branch to

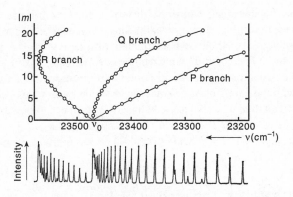

Fig. 7.10 The rotational structure in a $^1\Pi-^1\Sigma$ transition of a diatomic molecule. Three branches can be seen; the R branch forms a head because $B' < B''$.

close up slightly and those in the P branch to spread out with increasing J. These effects are much more exaggerated in an electronic transition, so that it commonly happens that the R branch spacing closes right down to zero; at this point, the branch turns round on itself and starts running back to lower wavenumbers. In the region where the branch turns round, there are several lines in close proximity. They are often not resolved, the lines piling up so that their intensities add to give a prominent feature in the spectrum called a *branch head*. Indeed, the branch head is usually the most obvious feature of a band even at low resolution.

The actual J value at which the R branch reaches the head depends on the relative values of the upper and lower state B values. From eqn 7.45, we have

$$d\nu_R/dJ = (3B' - B'') + 2(B' - B'')J. \qquad (7.46)$$

At the head, this derivative goes to zero, corresponding to

$$J_{\text{head}} = -(3B' - B'')/[2(B' - B'')]. \qquad (7.47)$$

An example of the rotational structure in a $^1\Pi - ^1\Sigma$ band of a diatomic molecule is shown in Fig. 7.10.

The situation which we have described above, with head formation in the R branch, is the one which occurs most commonly because it corrresponds to a longer bond in the upper electronic state. (Promotion of an electron from a lower to a higher orbital energy tends to lead to a weakening of the bond.) The rotational structure spreads out to lower wavenumber and is said to be *red degraded* (i.e. towards longer wavelength). However, the opposite situation can occur. When B' is greater than B'', the head is formed in the P branch at

$$J_{\text{head}} = (B' + B'')/[2(B' - B'')]. \qquad (7.48)$$

In this case, the vibrational bands are said to be violet (or blue) degraded.

In order to extract the full information content from the rotational structure in an electronic spectrum, it is necessary to analyse this structure. This is carried out in a manner similar to that described for vibration–rotation transitions in Chapter 5. The branches are identified and the rotational quantum numbers assigned, often using the method of combination differences, see Section 5.8. If the first line in each branch can be identified, it is usually possible to assign J values on that basis alone. The upper and lower state B values (and centrifugal distortion corrections if necessary) can be determined by fitting formulae like eqns 7.42 to the experimental data; alternatively, the method of combination differences can be used to give separate values for B' and B''. This method is particularly useful for a spectrum in which the rotational levels of the (usually) upper state are perturbed through mixing with the levels of a different, nearby electronic state. The upper state levels will not then be well described by eqn 7.36. It is still possible to obtain a reliable value for B'' in these circumstances because the combination difference $\Delta_2 F''(J) \equiv R(J-1) - P(J+1)$ is independent of the upper state levels.

7.6   Dissociation and predissociation

In Fig. 7.8, we have shown the potential energy curves for a typical electronic transition in a diatomic molecule; we call the ground state X and the excited state A. Let us consider a situation in which the molecules start in the $v'' = 0$ level and absorb radiation so that they are excited to a level v' in the upper A state, with a probability given by the appropriate Franck–Condon factor. It may well happen that the amount of energy absorbed from the radiation is larger than that required to dissociate the molecule into two atoms, D_0'', measured from $v'' = 0$. This dissociation will probably not occur to any significant extent because the transition to the bound vibrational level v' in the A state is much more likely. (The quantized levels of the upper electronic state are embedded in the continuum of levels corresponding to the two atoms A and B above the dissociation limit of the ground electronic state.) In Fig. 7.8, the most probable transition from the ground state is to the level $v' = 4$. From here, the excited molecule could re-emit the radiation and return directly to the X state by *fluorescence*. It could also lose vibrational energy by falling to the level $v' = 0$ by collisional relaxation; from here it can re-emit radiation and so return to ground state levels (though not necessarily $v'' = 0$). There is also a slight possibility that a *radiationless transition* will occur from the level $v' = 0$ to a higher vibrational level of the ground electronic state, $v'' = 11$; this process is much more likely to occur in the condensed phase than in a gas at low pressure.

Absorption to a higher energy, above the dissociation limit of the A state, does lead to dissociation because here there are no competing processes. One product of the dissociation in this case is an atom in an excited electronic state (A*) with excitation energy

$$E_{\text{ex}} = (D_0' + \nu_{00} - D_0'') \tag{7.49}$$

where ν_{00} is the difference in energy between the levels $v' = 0$ and $v'' = 0$, see Fig. 7.8. It is unusual but not impossible for this excess energy to be shared between the two atomic products. In a well-known example, the visible absorption spectrum of I_2 shows a set of converging bands which lead to a continuum corresponding to the dissociation of the upper $^3\Pi_{0_u^+}$ excited state into a ground state I atom, $^2P_{3/2}$, and an excited atom in the $^2P_{1/2}$ level. The ground state of I_2 dissociates into two ground state atoms, $^2P_{3/2}$, so that E_{ex} is equal in this case to the excitation energy $^2P_{1/2} \leftarrow {}^2P_{3/2}$, namely 7603 cm^{-1}.

The absorption spectrum in such a case consists of a series of sharp bands (actually the most prominent features for I_2 are the R branch heads) which

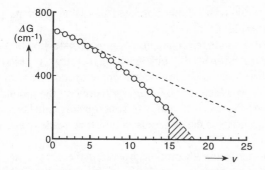

Fig. 7.11   The determination of the dissociation limit of the ultraviolet bands of O_2 from a Birge–Sponer extrapolation of the vibrational intervals, $\Delta G(v)$. The hashed area shows the energy difference between the series limit and the last observed band. Note the deviation from the linear extrapolation.

correspond to smaller values of v'. As v' increases, the bands get closer together, as expected from considerations of anharmonicity. Above the point where adjacent bands converge, the absorption is replaced by a continuum. This continuum corresponds to the arbitrary amount of energy taken away as kinetic energy of the two atomic fragments. The onset of this continuum gives an accurate measurement of the excited state dissociation energy referred to the $v'' = 0$ level of the ground state. If the atomic excitation can be identified, so that E_{ex} in eqn 7.49 is known, the ground state dissociation energy D_0'' can be deduced, since ν_{00} can be measured directly from the spectrum.

It often happens that the absorption near the dissociation limit is too weak to be observed so that the onset of the continuum cannot be identified directly. Its position can nevertheless be estimated by extrapolation from the v' levels lower down. We have seen in Section 5.6 that the spacing between adjacent vibrational levels of a diatomic molecule gets progressively smaller as v increases because of anharmonicity; at the dissociation limit, this separation goes to zero. If the lower vibrational levels are well represented by

$$G(v') = (v' + \tfrac{1}{2})\omega'_e - (v' + \tfrac{1}{2})^2 \omega_e x'_e; \qquad (7.50)$$

the separation $\Delta G(v')$ goes to zero at $(v'_L + \tfrac{1}{2}) = \omega'_e / 2\omega_e x_e'$ and the energy at this point is $(\omega'_e)^2 / 4\omega_e x_e'$ measured above the minimum of the A state potential energy curve. Thus, provided the vibrational energy levels can be adequately represented by the eqn 7.50, we can estimate the dissociation energy for the upper electronic state. This formula corresponds exactly to that for a Morse potential; more complicated energy expressions can be handled in a similar way. A graphical method based on eqn 7.50 is also often used. The intervals between successive band heads, i.e. $\Delta G(v')$, are plotted versus a running number and extrapolated linearly to zero. The area under this graph then gives the additional energy required to reach the dissociation limit. The method is known as the Birge–Sponer extrapolation. It is important to realize that this method of determining the dissociation energy of a molecule AB is, at very best, an estimate. It suffers from two weaknesses. First, as with any extrapolation, errors in the measurements of the band heads are exaggerated at the point where the dissociation limit is reached. Secondly, the Morse oscillator formula eqn 7.50 on which it is based is likely to become less reliable as v increases. Indeed, the behaviour of the energy levels near dissociation is likely to show a marked deviation from its predictions, as shown for example in Fig. 7.11.

In principle, a similar extrapolation of ground state vibrational intervals can also be used to estimate D_0'' but levels higher than $v'' = 0$ or 1 are not commonly observable in absorption at room temperature and high values of v'' are not always prominent in emission. Furthermore, the ground state curve is usually deeper, the anharmonicity is consequently smaller, and the results become less accurate as the extrapolation must be made over a greater energy range.

What happens in the absorption process if the upper state is not bound? Two different arrangements of potential energy curves are given in Fig. 7.12. The only possible outcome of the absorption process in these cases is dissociation, giving rise to a continuous spectrum without vibrational features. The upper states in Fig. 7.12 are purely repulsive with no stable minimum. In reality, they correspond to the potential energy of a non-associative collision between two atoms which experience a steep repulsion when they are approached to a distance less than the sum of their van der Waals' radii. The onset of the continuum, that is the lowest possible absorption energy, occurs at the energy required to form two atoms which may (Fig. 7.12(b)) or may not (Fig. 7.12(a)) be in their ground atomic states. The most intense part of the continuum corresponds, in accordance with the Franck–Condon principle, to vertical transitions such as those shown. These occur at considerably greater wavenumbers than the edge of the continuum; this edge is usually very ill-defined and difficult to identify accurately. Indeed, it may be more worthwhile to use some method of detecting the atoms rather than the absorption process; in this way, the smallest wavenumber required for the production of atoms may be detected with greater sensitivity. Observations of continua of the type shown in Fig. 7.12(a) are more common than those of the type shown in Fig. 7.12(b). Where two curves correspond to a common dissociation limit, they usually have different spin multiplicities; electronic transitions between them are then weak or forbidden. For example, the levels of H_2 are disposed as in Fig. 7.12(b), but the repulsive curve corresponds to a triplet state whereas the bound, ground state is a singlet.

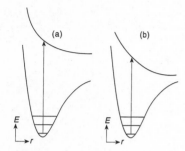

Fig. 7.12 Potential energy curves with repulsive upper states dissociating to (a) at least one excited atom and to (b) the ground state atoms.

Predissociation effects in electronic spectroscopy

Another interesting dissociation phenomenon is illustrated in Fig. 7.13. Here the upper electronic state A is bound but is crossed by a dissociative state B. To a first approximation, these states can be considered separately. If the electronic transition between the A and X states is allowed, a spectrum with a system of vibrational bands is expected. However, it is quite possible that small terms in the full molecular Hamiltonian will mix the A and B states and the potential energy curves, appropriate to fixed nuclei, follow the dashed curves near the crossing point. This is a manifestation of what is called the *non-crossing rule* (due to E. Teller). The vibrational levels below the crossing point ($v' \leq 4$ in Fig. 7.13) remain well quantized, and transitions to these levels give the usual sharp bands. So too do transitions to levels well above the crossing point. This is less expected at first sight, because there is more than enough energy to cause dissociation. However, if absorption of radiation leaves the system with energy nearly equal to that at the crossing point, the system is in a state which is a linear combination of wavefunctions appropriate

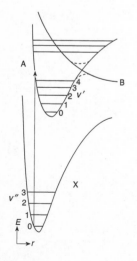

Fig. 7.13. Potential energy diagram showing predissociation of the vibrational levels of the upper, A state around $v' = 6$.

to curves A and B. The likelihood that two atoms will follow curve B out to dissociation is quite high. The lack of quantization of the relative kinetic energy of the two atoms means that there is no precise energy requirement and the absorption spectrum is continuous. Put another way, the lifetime of the molecule in the excited electronic state is of the order of a vibrational period (10^{-13} s). This is much too short a time for the molecule to execute a full rotation, and rotational motion ceases to be quantized. The rotational structure of the vibrational band in question is therefore washed out; in particular the rotational band head disappears and the vibrational band appears 'fuzzy'. The precise calculation of the probability of dissociation through, say, the level $v' = 6$ is quite difficult, but there is a characteristic that the probability is high if, expressed in classical language, the system has very little kinetic energy when it is near the crossing point. For this requirement, there is a good analogy with a straight putt on the golfing green: if the ball's kinetic energy is low when it reaches the hole, it will roll in but, if the velocity is too high, the ball crosses over the top of the hole and continues its path on the far side.

The resultant absorption in such a case consists of a progression of sharp vibrational bands which is smeared out for a section in the middle by the loss of rotational structure. The exact extent to which rotational (and even vibrational) structure is lost depends on the probability of crossing to state B. At higher wavenumbers, the bands converge to the normal type of dissociation continuum when the energy is above the asymptote to curve A. The dissociation which occurs at the crossing point with the repulsive curve B is known as *predissociation* because it occurs before the regular dissociation limit. It can be seen from Fig. 7.13 that the onset of predissociation in an absorption spectrum gives an upper limit for the dissociation energy of the ground electronic state, not a very good value perhaps but better than nothing!

8 Photoelectron spectroscopy

8.1 Introduction

We have seen in Chapter 7 how a molecule can fall apart if given enough energy in the form of radiation (photodissociation). If it is given a somewhat larger amount of energy, it can also be made to ionize

$$M + h\nu = M^+ + e; \qquad (8.1)$$

the process is known as *photoionization*. It is very similar to the photoelectric effect in which radiation of a variable frequency is shone on the surface of an easily ionizable metal (like an alkali). Provided the photon energy $h\nu$ is greater than a certain amount, known as the work function, electrons are emitted. The term photoionization is usually used to describe the same process when it is carried out in the gas phase. The threshold energy in this case is known as the *ionization energy*.

The energy required to ionize a hydrogen atom in a particular orbital n, l is just the energy required to remove that electron from the influence of the Coulombic attraction of the nucleus (i.e. to infinity). In other words, it is the energy of the electron in that orbital, measured relative to the ionization limit. We have seen in the previous chapter how a similar orbital description can be applied to molecules. In that case, however, the description is only approximate, requiring the repulsive interaction between electrons to be neglected. Nevertheless, it provides a guide to the ordering of electronic levels and allows their symmetries to be determined. In this picture, the ionization energy can be equated with the energy required to remove an electron from a particular molecular orbital of the molecule. The first ionization energy I_1 gives the energy to remove the electron from the highest occupied orbital in the neutral molecule, I_2 gives the energy required to remove an electron from the next highest orbital and so on.

If a gaseous sample of molecules M is irradiated with a beam of light of a single, fixed frequency ν_0, the excess energy (over the amount required to ionize the neutral molecule) appears almost completely as translational energy of the ejected electron. If the excess kinetic energy of the electrons is measured and subtracted from the energy of the photons $h\nu_0$, the difference is the ionization energy of the molecule. If there are several possible ionization processes which can occur below the photon energy, each will give a signal at the appropriate electron kinetic energy. This set of observations constitutes the photoelectron spectrum. The energy level scheme for the experiment is shown in Fig. 8.1.

The energy required for the experiment depends on the photoionization process involved. For example, the energy required to remove a valence electron is about $80\,000$ cm^{-1} or 10 eV. A photon of this energy falls in the ultraviolet region of the spectrum (strictly the vacuum ultraviolet) and so an experiment of this type is called ultraviolet photoelectron spectroscopy or UVPES (often abbreviated to UPS). On the other hand, a source of soft (or

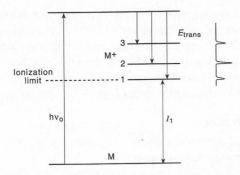

Fig. 8.1 The energy level scheme for photoelectron spectroscopy. When a sample of molecules M is irradiated with monochromatic light, the ionization energy can be determined by subtracting the excess energy, which is carried off as translational energy of the electrons, from the energy of the photon.

low energy) X-rays has enough energy (about 1000 eV) to remove inner-shell or core electrons. The technique in this case is referred to as X-ray PES or XPS. The first UPS experiments were carried out independently by F. I. Vilesov in Leningrad and by D. W. Turner in London in 1962; the original XPS experiments were performed a little earlier by K. Siegbahn in Sweden. In this chapter, we shall confine our attention to UPS.

8.2 Principles of photoelectron spectroscopy

The experiment

The basic arrangement of the photoelectron spectroscopy experiment is shown in Fig. 8.2. Monochromatic radiation in the vacuum UV is focused onto a gaseous sample of molecules M, causing them to be ionized according to eqn 8.1. The excess energy from this process, over and above the ionization energy, is released almost completely as kinetic (i.e. translational) energy of the electrons. The electrons are ejected in all directions; a small fraction passes through a slit into the analyser region. The voltage across the curved electrostatic analyser is such that only electrons of a given energy reach the detector; the voltage can be varied, thereby changing the kinetic energy required for transmission. The photoelectron spectrum is then recorded as the number of electrons reaching the detector per second as a function of their kinetic energy.

The monochromatic source of light is usually an atomic lamp. A commonly used one is based on the He I resonance line at 58.4 nm (21.22 eV) which corresponds to the $1s^1 2p^1 - 1s^2$ ($^1P - {}^1S$) transition. It was through the development of such a lamp that Turner was able to record the first photoelectron spectra. Since those days, VUV radiation has also been provided for PES by a synchrotron source and, in more recent times, by lasers.

The spectral linewidth achieved in photoelectron spectroscopy is rather poor compared with the other branches of spectroscopy described in this book. In practice, it is typically about 10 meV (or 80 cm^{-1}) although linewidths as small as 4 meV have been achieved in particular experiments. This means that the technique is limited to the resolution of vibrational structure only; it is not possible to observe rotational structure in a photoelectron spectrum. In the

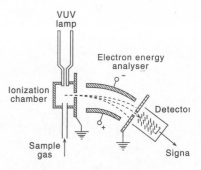

Fig. 8.2 The experimental arrangement for ultraviolet photoelectron spectroscopy. The energy analyser can be made to transmit only electrons of a particular kinetic energy by adjustment of the voltage across it.

VUV experiment, the linewidth is not limited by that of the radiation source but rather by instrumental factors such as the cleanliness of the analyser surfaces. A recent variant of the photoelectron spectroscopic experiment, based on tunable lasers and confined to the detection of electrons at threshold, has a much higher resolution (a linewidth of 1 cm^{-1} or less). This experiment is known as zero kinetic energy electron or ZEKE spectroscopy.

Energy considerations

If we apply monochromatic radiation of frequency ν_0 to our sample, the conservation of energy requires that

$$h\nu_0 = E_{ion} + E_{trans} - E_M \qquad (8.2)$$

(see Fig. 8.1). Although the energy of the parent molecule M is quite small (it is just the thermal energy), that of the photon is large. This energy is shared between the products of the photoionization process in eqn 8.1; the manner in which it is shared provides the information content of the photoelectron spectrum.

In addition to conforming to the conservation of energy, the photoionization process must also obey the conservation of linear momentum. The momentum of M and that of the photon on the left-hand side of eqn 8.1 are very small and can therefore be neglected (the photon has a momentum of h/λ from the de Broglie relation). Thus, if the ion M$^+$ and the photoelectron move apart with velocities $\mathbf{v}_{ion}$ and $\mathbf{v}_e$ we have

$$M_{ion}\mathbf{v}_{ion} + m_e\mathbf{v}_e \approx 0. \qquad (8.3)$$

Thus

$$\mathbf{v}_e \approx (M_{ion}/m_e)\mathbf{v}_{ion}. \qquad (8.4)$$

From this, it is easy to show that the kinetic energy of the electron is a factor (M_{ion}/m_e) larger than the kinetic energy of the ion M$^+$. The least favourable case we can consider is the photoionization of H$_2$; even for this, the kinetic energy of the electron is 3600 times larger than the ion kinetic energy. It is thus quite justifiable to assume that all the surplus kinetic energy in the process 8.1 is carried off by the electron. Rearranging eqn 8.2, we see that kinetic energy of the electron is given by

$$E_{trans} = h\nu_0 - (E_{ion} - E_M). \qquad (8.5)$$

The energy of the ion M$^+$ has a translational and an internal contribution; the same applies to the parent molecule M. From the considerations above, we see that the translational contributions to M$^+$ and M can be ignored. The internal contributions can be further partitioned in accord with the Born–Oppenheimer principle:

$$E_{int} = E_{el} + E_{vib} + E_{rot}. \qquad (8.6)$$

Thus the measured kinetic energy of the photoelectron gives the difference in energy between the ion and the parent neutral molecule.

Electronic energy contribution

The ionization process can be thought of, to a first approximation, as the excitation of M in its ground state to form M$^+$ in a different electronic state. In other words, the bulk of the energy given to M is used to increase the electronic energy of the system. This *difference in electronic energy* is known as the ionization energy (it is also called the ionization potential in some older textbooks).

If we ask ourselves what transitions of this type are allowed, it may be rather surprising to learn that there are almost no restrictions at all on this process. The selection rules governing photoionization are very lax (although it does turn out that one-electron excitations are much more likely to occur in practice). This is in fact one of the strengths of the experiment—it is possible to reach a much greater variety of electronic states of the ion than it is in conventional electronic spectroscopy described in Chapter 7. The one selection rule which applies to photoelectron spectroscopy concerns the allowed change in electron spin multiplicity. It is

$$\Delta(2S + 1) = \pm 1 \tag{8.7}$$

and results from the fact that the ejected electron carries off a spin angular momentum of $\frac{1}{2}$ in the process.

The molecular orbital description of electronic structure, introduced in the last chapter, provides a very simple way of understanding the photoionization process. Suppose that an electron is allocated to a particular orbital in the neutral molecule. Photoionization consists simply of the removal of the electron from this orbital to infinity, leaving all the other electrons undisturbed in their original orbitals. In other words, the energy I_i required to remove the electron is a direct measurement of its orbital energy

$$I_i \approx -\varepsilon_i \tag{8.8}$$

where ε_i is the orbital energy introduced in eqn 7.11. The negative sign is required because orbital energies are measured relative to the ionization limit and so are defined to be negative. The interpretation in eqn 8.8 was first proposed by Koopmans and is known as Koopmans' theorem. (Strictly speaking, the orbital energy in eqn 8.8 should be interpreted as the self-consistent field or SCF energy. In an SCF calculation, some account is taken of the presence of the other electrons in the molecule. Also, Koopmans' theorem applies only to closed-shell molecules.)

We know from its dependence on molecular orbitals that Koopmans' theorem must be approximate. Molecular orbital theory assumes that the electrons in a molecule move independently and so ignores the correlation between them caused by electron–electron repulsion. In addition to this, Koopmans' theorem assumes that the orbitals of M and M^+ are identical. This is clearly not the case because the electrons which remain in M^+ will rearrange themselves slightly after the ionization to establish a new minimum of energy (a phenomenon which is called electron reorganization).

The quantity which is measured in photoelectron spectroscopy is the ionization energy which corresponds to the difference in energy between the ion M^+ and its neutral parent M; this quantity is accurately defined. The approximation in Koopmans' theorem comes from the involvement of orbitals which are, in the final analysis, an imperfect theoretical construct. Despite this, the theorem is very useful because it shows that the ionization energies measured in a photoelectron spectrum give a direct representation of the molecular orbital diagram of the molecule, in so far as it is valid. It is this simple interpretation of a photoelectron spectrum which is so appealing to physical chemists.

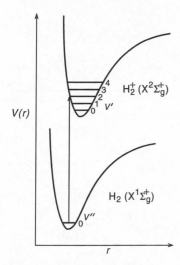

Fig. 8.3 The potential energy curves for the photoionization of H_2. The ion H_2^+ is less strongly bound than H_2 and so has a longer bond length. Consequently, the most favoured vibrational transition from the ground state of H_2 is to the level $v' = 2$ of the X state of H_2^+.

Vibrational structure in photoelectron spectroscopy

We know from earlier chapters that the vibrational motion of a molecule can make a significant contribution to its internal energy, eqn 8.6. In the photoelectron spectroscopy experiment, we usually start with the neutral molecule M in its zero point vibrational level. However, excitation can produce the ion M^+ in one of several vibrational levels of the upper electronic state. To the extent that we can separate electronic and vibrational motion, the relative intensities of these different vibrational bands are governed by the Franck–Condon principle, exactly as they are for electronic spectroscopy (section 7.5). Thus, for example, if the electron is removed from a non-bonding orbital of M, the potential energy curve and the r_e value for M^+ will be very similar to those of the parent molecule. Thus, the bulk of the transition intensity is confined to the $(v', v'') = (0, 0)$ band and there is essentially no vibrational structure in the photoelectron spectrum. However, the removal of an electron from a bonding or anti-bonding orbital causes a large change in bond length and consequently extensive vibrational structure in the photoelectron spectrum; there would be a long vibrational progression as shown in Fig. 7.9. Such behaviour is illustrated nicely by the photoelectron spectrum of H_2. The two electrons in the single bond in this molecule are in the $1\sigma_g$ (1s) orbital (see Fig. 7.1). Removal of one of these electrons leads to a reduction in the binding energy and a longer bond length r_e in the $X^2\Sigma_g^+$ state of H_2^+ which is formed, see Fig. 8.3. The most probable vibrational transition is to the level $v' = 2$ by the Franck–Condon principle. The long progression which results is shown clearly in the photoelectron spectrum of H_2 in Fig. 8.4.

The difference in the electronic energies of M and M^+ is obviously best defined as that corresponding to the (0,0) band. This quantity is known as the *adiabatic ionization energy*. However, it may not be very easy to measure this quantity in practice, especially if the vibrational progression is so long that the first members are weak and not observed. Because of this, a quantity known as the *vertical ionization energy* is defined in practice. This is the energy which corresponds to the intensity maximum in the $v'' = 0$ progression. We note that,

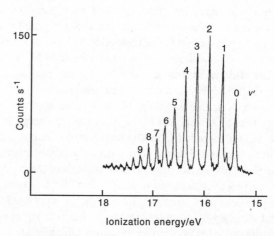

Fig. 8.4 The UV photoelectron spectrum of the H_2 molecule recorded with He I radiation (21.22 eV).

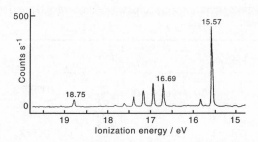

Fig. 8.5 The UV photoelectron spectrum of the N_2 molecule recorded with He I radiation (21.22 eV). Three separate ionization processes can be discerned, corresponding to transitions to the X, A and B states of N_2^+.

since Koopmans' theorem makes no allowance for the reorganization of electrons or nuclei, it is probably more appropriate to interpret it in terms of the vertical rather than the adiabatic ionization energy. Such ambiguities of interpretation remind us not to place too much trust in the quantitative predictions of Koopmans' theorem.

Rotational structure

In conventional photoelectron spectroscopy, it is very rare for the rotational structure in a vibrational band to be resolved. However, the laser techniques employed in ZEKE spectroscopy allow spectra to be recorded with much higher resolution and rotational structure is routinely observed, even for quite large molecules such as benzene. The rotational selection rules in this case are not very restrictive because they depend on the angular momentum carried off by the ejected electron. This can be seen by applying the conservation of angular momentum to the overall process (eqn 8.1); the laxness of the ΔJ selection rule arises because the photoelectron can take away any amount of angular momentum.

8.3 Example of a photoelectron spectrum: molecular nitrogen

The UV photoelectron spectrum of N_2 recorded with He I radiation is given in Fig. 8.5. The ground electron configuration of N_2, which was discussed in the previous chapter, is $...(2\sigma_g)^2 \ (2\bar{\sigma}_u)^2 \ (1\pi_u)^4 \ (3\sigma_g)^2$; the corresponding molecular orbital diagram is given in Fig. 8.6. We see that the lowest ionization energy (15.57 eV) corresponds to removal of an electron from the outermost $3\sigma_g$ orbital. The second (16.69 eV) and third (18.75 eV) ionization energies correspond to removal of an electron from the $1\pi_u$ and $2\bar{\sigma}_u$ orbitals respectively. The electronic states of N_2^+ formed by these three processes are $X^2\Sigma_g^+$, $A^2\Pi_u$ and $B^2\Sigma_u^+$ in order of increasing energy.

According to the discussion of the molecular orbital diagram in section 7.2, we expect the $3\sigma_g$ and $1\pi_u$ orbitals to have bonding character (the latter more so than the former) and the $2\bar{\sigma}_u$ orbital to have antibonding character. Looking at the photoelectron spectrum in Fig. 8.5, we see that there is a long vibrational progression associated with the ionization I_2, which is indeed consistent with the bonding character of the $1\pi_u$ orbital. However, the progressions associated with the ionizations I_1 and I_3 are both short, which implies that the bonding and

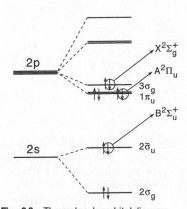

Fig. 8.6 The molecular orbital diagram for the N_2 molecule, showing the ground electronic configuration. Removal of an electron from the highest occupied orbitals produces the three ionizations observed in the photoelectron spectrum shown in Fig. 8.5.

Table 8.1 Bond lengths and vibrational wavenumbers of N_2^+ and N_2 in various electronic states

molecule	configuration	state	r_e/nm	ω_e/cm^{-1}
N_2	$(2\bar{\sigma}_u)^2 (1\pi_u)^4 (3\sigma_g)^2$	$X^1\Sigma_g^+$	0.109769	2358.6
N_2^+	$(2\bar{\sigma}_u)^2 (1\pi_u)^4 (3\sigma_g)^1$	$X^2\Sigma_g^+$	0.111642	2207.0
N_2^+	$(2\bar{\sigma}_u)^2 (1\pi_u)^3 (3\sigma_g)^2$	$A^2\Pi_u^+$	0.11749	1903.7
N_2^+	$(2\bar{\sigma}_u)^1 (1\pi_u)^4 (3\sigma_g)^2$	$B^2\Sigma_u^+$	0.10749	2419.8

antibonding characteristics of the $3\sigma_g$ and $2\bar{\sigma}_u$ orbitals are, rather unexpectedly, only slight. Careful scrutiny of the spectrum also shows that the vibrational spacings are different in the three transitions. In particular, the interval for the I_2 ionization to the $A^2\Pi_u$ state is 1904 cm^{-1}, significantly smaller than the vibrational interval in the parent molecule (2359 cm^{-1}). This is also indicative of the bonding character of the $1\pi_u$ orbital. The interpretation of the photoelectron spectrum (Fig. 8.5) is also consistent with the bond lengths of N_2^+ in the three electronic states, which have been determined by high-resolution electronic spectroscopy. These bond lengths are given, together with the harmonic vibrational wavenumbers, in Table 8.1. The largest change in bond length and vibrational wavenumber occurs on excitation from N_2 to the A state of N_2^+. In the X state of N_2^+, the bond length is slightly longer and for the B state it is slightly shorter than for N_2 consistent with weak bonding and antibonding character of the $3\sigma_g$ and $2\bar{\sigma}_u$ orbitals.

8.4 Autoionization

The direct photoionization process given in eqn 8.1 takes the molecular system from a region of bound, quantized energy levels to a continuum above the ionization limit. There is another, indirect route to the same final products, M^+ + e; this process is known as *autoionization*. It is summarized in Fig. 8.7. The molecule M is first excited to an electronic state of the neutral molecule with energy above the ionization limit. From here, it can emit an electron spontaneously to form the ion M^+:

$$M + h\nu \rightarrow M^* \rightarrow M^+ + e \tag{8.9}$$

The first step of this scheme is a resonant process and obeys the usual electric dipole selection rules; it can be brought about only by light of the appropriate wavelength (or energy). The autoionization process is the second step. Since no radiation is involved, the only requirement is that the symmetry of the autoionizing state must be the same as that of the final state plus the electron. The overall process, eqn 8.9, must satisfy the conservation of energy. Consequently, the kinetic energy of the electrons ejected in the second step is the same as the photoelectrons would have if they were ionized directly with light of the same wavelength. However, the probabilities of the direct and indirect processes are often very different. As a result, the intensities of the different electronic transitions and their vibrational structure are also different.

Autoionization will not affect a photoelectron spectrum unless it so happens that the frequency of the ionizing radiation just matches a transition

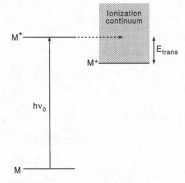

Fig. 8.7 The process of autoionization. The molecule M is resonantly excited to a level of a high lying, Rydberg state (M*) which lies above the ionization limit. From here, the molecule can expel an electron to produce the ion; the electron carries off the excess energy as translation.

frequency of the neutral molecule. This occurs rarely but, when it does, it can produce a spectrum with a markedly different appearance. The autoionizing state is generally a Rydberg state of M, belonging to a series which converges on some higher ionization limit (the state must obviously lie below this limit). Because the ionization energy of most valence electrons is less than 21 eV, it is very unusual to observe autoionizing states with He I radiation. However, if a lower energy atomic lamp, such as the Ne I resonance line at 73.6 nm (16.85 eV), is used, several autoionizing states can be encountered. There is a well-known autoionization in molecular oxygen at this wavelength, for example.

8.5 Zero kinetic energy (ZEKE) spectroscopy

The resolution in conventional photoelectron spectroscopy is limited by the accuracy with which the kinetic energy of the electrons can be measured; it is typically 10 meV (80 cm^{-1}). As a result, much structural information, notably rotational structure, is not accessible to this experiment. A variant of photoelectron spectroscopy which has a much higher resolution has recently been developed by K. Müller-Dethlefs and colleagues in Germany. The technique is called zero kinetic energy photoelectron spectroscopy or ZEKE spectroscopy. The linewidth achieved in these experiments can be as small as 0.2 cm^{-1}, providing an improvement in resolution of over two orders of magnitude.

In the conventional photoelectron experiment, shown in Fig. 8.1, the incident radiation has much more energy than is necessary to ionize the neutral molecule. The excess energy is carried off as kinetic energy of the photoelectron. If, on the other hand, we have available a *tunable* source of radiation, we can adjust the frequency so that it provides just enough energy to excite the molecule to a particular rotational–vibrational level of a given electronic state of the ion M$^+$. In this case, the photoelectrons appear with zero kinetic energy. If we arrange to detect just those electrons which have zero kinetic energy, we can record a spectral line whose width is dictated only by that of the radiation source. The feasibility of such an experiment depends largely on the development of narrow-line, tunable sources of coherent radiation in the vacuum UV or at even shorter wavelengths. The radiation is usually provided by lasers whose frequency has been doubled or tripled.

The way in which the signal at the detector can be limited to those electrons with zero kinetic energy is based on a very simple idea. The molecules M in the sample chamber are ionized with a pulse of radiation of the appropriate frequency. Photoelectrons are ejected in all directions and with a spread of energy. The electrons with non-zero kinetic energy move away from the photoionization region whereas those with zero kinetic energy remain where they are formed. After a short time interval, therefore, only the electrons in which we are interested remain; they have become spatially separated from the moving electrons. These electrons can then be extracted into a time-of-flight drift tube by applying a pulsed electric field and will arrive at the detector after a characteristic, determinable delay to constitute the signal. The ZEKE spectrum consists of the electron signal plotted as a function of the frequency of the incident radiation.

The precise definition of the zero kinetic energy electrons can be somewhat spoiled in practice by small, stray electric fields which inevitably occur in the ionization region. A refinement of the basic ZEKE experiment has therefore been developed. Tunable radiation is applied to the molecule M with energy very slightly less than the ionization energy. This excites the molecule to a very high Rydberg state in which an electron is orbiting at a large distance around the M^+ ion core. This electron is very loosely held by the molecule and the Rydberg state can be easily ionized by application of a small, pulsed electric field which pulls the electron away from the core. As before, there is a delay between the laser and electric field pulses. During this interval, the M* molecules do not have time to move away and the field-ionized electrons are formed at the original excitation site (i.e. in the same place as the ZEKE electrons). Since they are formed only on application of the electric field, they can be detected much more efficiently than the ZEKE electrons. The technique is known as ZEKE-PFI (pulsed field ionization). While PFI and ZEKE electrons give rise to almost identical spectra, there is a small energy shift between them. This difference can be accounted for empirically to give the true, zero-field ionization energies.

ZEKE spectroscopy has two advantages over conventional photoelectron spectroscopy. As already mentioned, the resolution is much higher and it is routinely possible to observe rotational structure. For example, the ZEKE spectrum of the $X^1\Sigma^+ - B^2\Sigma^+$ transition of nitric oxide (NO) has been recorded at rotational resolution; a small part is shown in Fig. 8.8. In addition, because ZEKE spectroscopy is based on pulsed lasers, measurements can be made as a function of time and it is possible to study the *dynamics* of the various processes involved.

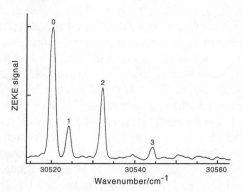

Fig. 8.8   Part of the ZEKE spectrum of the $X^1\Sigma^+ - B^2\Sigma^+$ ionization of nitric oxide. The lines correspond to individual rotational transitions from the level $N'' = 0$ in the B state. The number beside each line is the value of the rotational quantum number N' in the upper state (that of the ion). Note that changes in N of up to 3 can be observed.

Bibliography

Atkins, P. W. and Friedman, R.S. (1997). *Molecular Quantum Mechanics*, 3rd edn, Oxford University Press.

Banwell, C. N. and McCash, E. M. (1994). *Fundamentals of Molecular Spectroscopy*, 4th edn, McGraw–Hill, London.

Barrow, G. M. (1962). *Introduction to Molecular Spectroscopy*, McGraw-Hill, New York.

Bernath, P. F. (1995). *Spectra of Atoms and Molecules*, Oxford University Press.

Bunker, P. R. (1979). *Molecular Symmetry and Spectroscopy*, Academic Press, London.

Dixon, R. N. (1965). *Spectroscopy and Structure*, Methuen, London.

Eland, J. H. D. (1984). *Photoelectron Spectroscopy*, 2nd edn, Butterworths, London.

Hollas, J. M. (1996). *Modern Spectroscopy*, 3rd edn, John Wiley and Sons, Chichester, UK.

Hollas, J. M. (1982). *High Resolution Spectroscopy*, Butterworths, London. Reprinted as a 2nd edn by John Wiley and Sons, Chichester, 1998.

Herzberg, G. (1950). *Spectra of Diatomic Molecules*, 2nd edn, Van Nostrand, New York. Reprinted 1989 with corrections by Krieger Publishing Co., Malabar, Florida.

Kroto, H. W. (1975). *Molecular Rotation Spectroscopy*, John Wiley and Sons, London. Reprinted 1992 as a Dover Paperback edition.

Lefebvre-Brion, H. and Field, R. W. (1986). *Perturbations in the Spectra of Diatomic Molecules*, Academic Press, San Diego. A new edition of this book, re-titled *Spectroscopy and Dynamics of Diatomic Molecules*, is due to be published in late 1999 or early 2000, again by Academic Press.

Softley, T. P. (1994). *Atomic Spectra*, Oxford University Press.

Index